Vegetation in
Civil and Landscape
Engineering

VEGETATION IN CIVIL AND LANDSCAPE ENGINEERING

D.H. Bache

Department of Civil Engineering
University of Strathclyde, Glasgow

and

I.A. MacAskill

Consulting Engineer, Glasgow

GRANADA
London Toronto Sydney New York

Granada Technical Books
Granada Publishing Ltd
8 Grafton Street, London W1X 3LA

First published in Great Britain 1984 by Granada Publishing

British Library Cataloguing in Publication Data
Bache, D.H.
   Vegetation in civil and landscape engineering.
   1. Land use — Planning
   I. Title   II. MacAskill, I.A.
   333.7   HD111

ISBN 0-246-11507-6

Typeset by Columns of Reading
Printed and bound in Great Britain by
Mackays of Chatham, Kent

Contents

Preface

Land use and landscape are continuously changing — fashioned by natural processes and reflecting the aspirations of mankind. Considerable demands are placed on land in terms of agricultural output, mineral production, water resource development and recreational uses; or simply for habitation, industry and transport. In nearly all cases the pattern of land use is altered with forest lands converted to grazing, grazing in turn to arable, and each yielding to urban development. Concurrently we witness a greater incidence of flooding, accelerated erosion, desertification and many other maladies stemming from a poor understanding of effects of altered land use, and from plain mismanagement. Of significance are the hydrological consequences of land disturbance — and not surprisingly, since a principal path of surface water movement is via the labyrinth of vegetation covering the land surface. Less obvious, but often far reaching, are the associated ecological changes affecting flora and fauna, and thereby the lifestyle of dependent populations.

Many problems can be avoided by good planning, but this depends on the perception and cooperation between parties engaged in landscape design. Major projects such as highway construction often call for a broad range of expertise covering ecology, soil science, meteorology, civil engineering, landscape architecture and others. Each discipline requires at least an elementary understanding of the others in order that problems are understood and for compatible solutions to emerge. Where smaller projects are being considered, the number of disciplines involved is likely to be reduced, with civil engineers and landscape architects carrying the major responsibilities for design. In these situations the engineer's role is often hampered (unwittingly) by an undue emphasis, in his early training, on the use of inert materials.

This book has a number of objectives: first, it demonstrates the use of vegetation as an engineering medium and evaluates its role in environmental control. Second, it brings together a large body of information, hitherto uncollated, and acts as a guide to specialised literature. Third, it has an important educational role — introducing and demonstrating the use of scientific methodology for specifying the attributes of vegetation.

The text contains many design examples to demonstrate the processes of qualitative and quantitative assessment. It should be stressed, however, that the book makes no pretence of being a comprehensive design manual. Where the design processes require a deeper understanding of the factors involved, the reader is referred to the specialised literature. Also, when analysing the functional role of plants it is vital to consider other contributory factors — thus a holistic approach is essential; for many of the case situations examined, it has been necessary to digress from the particular subject of plants in order to satisfy this basic requirement.

Following a basic introduction to aspects of plants and soils considered in the later text, the reader is introduced to the role of plants in surface hydrology since this provides a common thread to environmental problems in many disciplines. Problems are varied, ranging from catchment water yields to erosion control and irrigation planning. Erosion control and soil stabilisation form the basis of a second major chapter which again dwells on problems associated with water transfer. Plants are widely used by landscape architects and others for functional purposes as well as the aesthetic. Roads, fields and dwellings can be partially protected from wind and snow by shelterbelts. These and other important areas are assessed in terms of design and analysis.

A final chapter aims to broaden insight into matters of environmental concern.

The content of the book is wide-ranging, covers many disciplines, and could not have been written without the considerable support of many individuals. The authors are particularly indebted to Dr A.G. Chantler and Dr G. Fleming for their advice during the preparation of Chapter 2 and to Dr R.C. Kirkwood for clarifying many botanical queries. Special thanks are due to Dr J.M. Caborn for so kindly reading and commenting on Chapter 5 and to Dr B.W. Bache for his discerning criticism and advice during the preparation of Chapters 1 and 4.

Thanks are due to Evelyn MacKinnon for the considerable task of deciphering bad hand writing and typing the manuscript.

Our wives and families have been dragged over waste tips and quarries, and have tramped miles of roads, rivers and coast in the cause of the book; in the final stages of preparation they have had to put up with a great deal. Without their forbearance, together with editorial and general support, the book would still be a distant prospect.

D.H.B., I.A.M.

1 Plants and soil

1.1 Introduction

Whether contemplating the intricacies of hydrology, the land disposal of sewage sludges or techniques for combating erosion, we face the problem of evaluating, and perhaps exploiting, the properties of the soil-plant system. This will be familiar territory to landscape architects, foresters and many other disciplines concerned with agronomy. Civil engineers, however, though receiving a basic training in soil mechanics, rarely study the basic aspects of soil fertility. When it comes to the actual plants, it is probable that the engineer will know little beyond an intuitive grasp gained from school biology or experiences derived from the garden. Yet, if one wants to have a meaningful discussion of the utility of plants in the engineering environment, it is essential to gain a basic appreciation of the soil-plant system as a medium. Plants and soils are an intimately connected system and in the present context the terms cannot be discussed in isolation. Even if we are mainly interested in the aerial parts of the plant we should always remember the countryman's adage, 'Ahrr, to be sure, the answer lies in the soil'!

In the following discussion, we attempt to introduce the reader to some basic ingredients of the soil-plant system. The presentation makes no pretence at being exhaustive or uniform, but rather serves to support the subsequent text. It is also directed at readers possessing a general background in science, but without specific knowledge of plant biology and soil science. Particular attention is paid to plant-water relationships as these pervade so many aspects of land surface hydrology, and are closely tied to many problems in environmental control.

1.2 Plant structure

The different parts of a plant's structure possess distinctive physio-logical functions but also offer other characteristics (see table 1.1) which may be of advantage to the land-use planner. For example, root-binding stabilises the soil and reduces erosion.

Table 1.1   Plant components, their function and related attributes

	Primary function	*Related attributes*
Roots	Anchorage Absorption Conduction Storage	Soil reinforcement Soil permeability
Aerial	Stems: Support Conduction Production of new living tissue Leaves: Photosynthesis Transpiration	Interception Shelter Storage Aesthetic

Though the internal structure of a plant may not appear to be particularly relevant to the interests of the practising engineer, it contains many features which are necessary to a proper understand-ing of the role of plants in surface hydrology and their interaction with the soil system. Useful background information on introductory plant biology may be obtained from texts such as Weier *et al.* (1974).

1.2.1 Roots
The functions of the root system are absorption, anchorage, conduc-tion and storage (Weier *et al.* 1974).

In land plants, large amounts of water are absorbed by roots and most plants obtain the bulk of the mineral salts they require via the roots. These, and sometimes together with stored foods, are trans-mitted to stems and leaves above ground. In reverse, foods manufac-tured in the leaves are conducted to the roots and support their growth (fig. 1.1).

The principal factors determining root development are: (a) nutrient availability both from the aerial parts of the plant and from the soil; (b) oxygen supply; (c) soil moisture content and its osmotic pressure; (d) soil temperature; (e) toxin and pathogenic levels in the soil; (f) the system of soil pores into which the roots can grow; and (g) the shear strength and compressibility of the soil. Of these, it is worth stressing that the roots of most crops can only function actively if they have an adequate oxygen supply. For most agricultural crops (once they have passed the seedling stage) the source of oxygen is soil air (Russell, 1973).

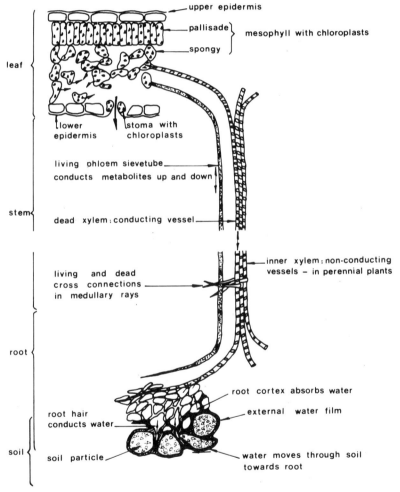

Fig. 1.1 Diagrammatic representation of the 'plumbing' system of a vascular plant

There are two general types of root system, a fibrous root system and a tap root system. The fibrous root system is characteristic of many plants, e.g. grasses, and is best adapted to shallow soils or to regions of light rainfall where moisture may not penetrate the soil deeply. The tap root system is well adapted to deep soils and to those in which the water table is relatively low. Most tap root systems are woody, but some like carrots are fleshy storage organs.

Plant roots are an important agent in promoting soil structure in undisturbed soils (section 1.5.1); this can result from root swelling which causes local consolidation, and enlarged pore spaces, which remain for a period after the root decomposes. Roots and their associated microbial populations also exude gums which stabilise aggregates in their vicinity. Humus formed from the decay of dead roots is also important in stabilising structure. The resulting crumb structure improves permeability and aids gaseous transfer. The efficiency of a grass cover in promoting aggregation is considered to be due to its rapid and prolific root production (Etherington, 1975). Plant roots also reinforce the soil, influencing the stability of the surface; these aspects are examined in chapter 3.

1.2.2 Stems and foliage

Stems support the photosynthetic and reproductive organs and provide a pathway for the movement of water and mineral nutrients from roots to leaves, and for the transfer of foods and other products essential to the metabolism of the plant.

The structure of the leaf shown in fig. 1.1 reflects its all-important function of photosynthesis. On the upper surface, immediately below the epidermis, is a group of closely packed cells containing chloroplasts (the palisade mesophyll). Below this is a zone of loosely packed cells (the spongy mesophyll). The intercellular spaces of both these tissues are gas-filled and open into larger cavities beneath the stomatal pores which connect with the atmosphere outside. Each stoma consists of a pair of elongated guard cells, occasionally supplemented by subsidiary cells as in grasses. The shape of the guard cells is such that when they are turgid (swollen), their adjacent walls separate to form a pore. The shape and width of the pore determines its performance in permitting the passage of water vapour and gases. The interaction between stomatal aperture and transpiration is discussed in sections 1.3, 1.4.4 and 2.2.4. The pore aperture diminishes proportionally with a reduction in turgor pressure in the guard cells; when they are flaccid (limp) the stoma are fully closed.

1.3 Water potential

Water in the soil-plant-air system is subject to many differing kinds of forces and it is desirable to express these in the same units wherever possible. This can be achieved using the concept of water potential; it is a thermodynamic variable expressing the free energy status of water and is measured in units equivalent to pressure.

Water transfer can only take place from regions of high potential to regions of low potential, i.e. it moves 'down' the 'pressure' gradient. For the soil-plant-air continuum, fig. 1.2 shows that the water potential is greatest in the soil and least in the atmosphere; this accounts for the upward transport of water from the soil into the atmosphere, either directly or via plants.

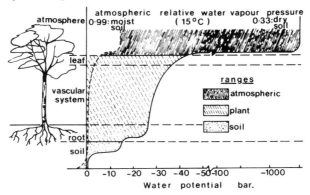

Fig. 1.2 A hypothetical distribution of water potential under different conditions of soil wetness and atmospheric water vapour pressure

The total water potential ψ_t at any position within the plant or soil is subject to contributions from gravity, pressure, and solute concentration. When the soil is saturated ψ_t is generally taken as positive, and becomes negative in unsaturated conditions. A negative water potential implies that water is being held by 'suction forces'; these can arise from surface tension forces, adsorption, or from the effects of solutes.

The concept of water potential is of considerable value when discussing the status of water and its movement through the soil-plant system. For present purposes we shall merely acknowledge its existence. For deeper insight, the reader is referred to Appendix 1 and to texts such as Marshall and Holmes (1979), Meidner and Sheriff (1976) and Kramer (1969).

1.4 Plants and water

1.4.1 The role of water in plants

Water is an all-pervading substance in much of the physical world and all of the plant world. Its importance to plants is represented by functions that can be grouped under four general headings (see Kramer, 1969) and discussed further in Meidner and Sheriff (1976).

(a) A constituent of protoplasm: Water constitutes 80-90% of the fresh weight of most herbaceous plants and over 50% of the fresh weight of woody plants.

(b) Solvent: A second essential function is that water serves as a solvent in which gases, minerals and other solutes enter plant cells. Most plant cells are permeable and a continuous liquid phase extends throughout much of the plant and enables the translocation of solutes.

(c) Reagent: Water is an essential ingredient of photosynthesis and is a reactant or reagent in many other important processes.

(d) Maintenance of plant turgor: Turgor pressure, i.e. the internal water pressure, plays an essential role in the control of plant metabolism and is capable of giving a high degree of rigidity both to the cell contents and the enveloping cell wall. In herbaceous plants this represents a major skeletal force providing support for the stems.

1.4.2 Entry of water into the plant

Reference to Appendix 1 shows that the absorption of water by plants is controlled by two groups of factors: those which affect the difference in the water potential from soil to roots; and those which affect the resistance to water movement through the soil and in the roots. Though the osmotic pressure within root cells normally exceeds that in the soil (so that water tends to enter the plant), it is generally considered that more water enters in response to the pressure potential (Winter, 1974).

EFFECT OF COMPOSITION OF SOIL SOLUTION

Usually solute levels in the cell sap are greater than in the soil solution and aid the inward flow of water. If the solute concentration in the soil water exceeds that within the cell sap, the flow may be reduced. Osmotic potentials of −2 to −4 bar at the permanent wilting point (discussed later) inhibit the growth of most plants, and soils with osmotic potentials approaching −40 bar are usually barren of vegetation (Kramer, 1969). Problems arising from osmosis are prevalent in arid and irrigated soils where evaporation concentrates solutes in the soil water. Similar effects can be induced by high concentrations of inorganic fertilisers.

EFFECT OF FLOODING
Gaseous transfer at the roots is inhibited by a waterlogged soil. Inadequate aeration inhibits metabolic activity of root cells which has the effect of reducing the permeability of root tissue to water (increased root resistance) and thus decreases absorption; this results in water deficit and even wilting of the shoots.

1.4.3 Water movement in the plant
The flow of water through the plant is largely a response to a pressure gradient as shown in fig. 1.2 and equation [A1.4]. Water potential within the leaf tissue is reduced by evaporation and exceeded by the water potential at the root surfaces giving rise to a pressure difference. The rate of flow is constrained by the resistance to flow in the roots, stem and leaf tissues.

1.4.4 Transpiration
The word transpiration refers to the loss of water vapour from plants. It is usually discussed in relation to leaves which not only form the major surface area of the shoots of most plants, but are also adapted to provide a pathway for gaseous exchange between the internal tissues and the atmosphere. The uptake of CO_2 by leaves is fundamental to the life of plants, and the loss of water vapour along the reverse pathway, though partially controlled, is largely inevitable.

Most water is lost from the mesophyll cells, from which water evaporates into intercellular spaces, and diffuses to the exterior via the sub-stomatal cavities and the stomatal pores. The important point here is that evaporation takes place *within* the plant tissue.

Reduction of CO_2 below atmospheric concentration causes stomatal opening. This is of major importance in causing opening in day-time, because light induces reductions of CO_2 concentration in and around the leaves through photosynthesis. The degree of opening is related to the light intensity, and at night the stomata are closed, i.e. transpiration ceases.

Stomatal closure can be induced by any factor which causes loss of turgor, e.g. high winds or reduced soil water content. However the most important generalisation about stomata is that they are normally open during the day and closed at night.

Stomatal behaviour and its influence on transpiration plays a key role in the understanding of evapotranspiration (sections 2.2.3 and

2.2.4) and is particularly important when considering water losses during the interception process (section 2.2.7) and during periods of substantial soil moisture deficit.

1.4.5 Plant growth and soil moisture
Either an excess or a deficiency of soil water limits root growth and its functions.

WATER STRESS

Plant water stress or water deficit refers to situations where cells and tissues are less than fully turgid. In simplest terms, water stress occurs whenever the loss of water by transpiration exceeds the rate of absorption. It is characterised by a decrease in water content accompanied by a lack of turgor, closure of stomata and decrease in growth. Little or no growth occurs in soils with a water content near the permanent wilting point (see section 2.2.5). This occurs when the soil water potential is equal to, or lower than the plant water potential, i.e. the plant cannot gain water unless the soil water potential is raised. The wilting point is usually taken as $\psi_{soil} = -15$ bar (pF 4.2, see Appendix 1 for definition of pF); this value is largely constant from soil to soil, but corresponds to different ranges of percentage water (see table 2.3).

EXCESSIVE SOIL MOISTURE

It is well recognised that the problem of excessive moisture as it affects growth is one which centres around deficient aeration. Water filling the soil pores not only displaces air, but also obstructs gaseous diffusion. Roots need oxygen; a root subjected to an oxygen shortage has its growth decreased and absorption is also reduced. Poor aeration also affects root growth and functions as a result of the products formed during anaerobic decomposition of organic matter by soil microorganisms (Russell, 1973). Waterlogging generally leads to a deceleration in the rate of organic matter decomposition — which means that organic matter (i.e. plant residue) accumulates after waterlogging. A consequence is that nitrogen tends to remain locked up in the organic residues, and nitrogen is often a limiting factor to plant growth on poorly-drained soils (van't Woudt and Hagan, 1957).

OPTIMUM SOIL MOISTURE

For most annual crop plants, water supply is optimal when the soil is at the 'field capacity' moisture content, that is when excess water has drained away under gravity after the soil has been nearly saturated.

In well-aggregated soils, field capacity corresponds to a soil moisture suction (= negative potential) of about 0.05 bar (pF 1.7); about 75-80% of the pore space is then occupied by water. The soil dries as roots extract water and the hydraulic conductivity is rapidly reduced, so that water stress occurs well before the permanent wilting point. Irrigation research indicates that for optimum growth the soil water should be maintained at 50-85% of the available water capacity.

THE WATER TABLE

Adequate crop production and the perpetuation of soil fertility in irrigated areas requires water table depths of 1.5 m or more (Hansen *et al.* 1980), but the optimum depth depends on particular crops under given conditions (van't Woudt and Hagan, 1957). A water table at 10 m or more is beyond the reach of most normal plants. The nearer a water table is to the surface, the greater the risk of it being harmful instead of beneficial. For example, if the water table is close to the surface only a small volume of soil is available for exploitation by the root systems and this increases the risk to drought should the water table fall temporarily (Winter, 1974).

1.5 Soil character

The ability of a soil to sustain plant growth and plant ecosystems depends on its ability to supply nutrients, water and air. Physically, a mineral soil is a porous mixture of inorganic particles, decaying organic matter, air and water as indicated in fig. 1.3. Two important physical characteristics which will be considered are soil texture and structure. Soil chemical properties give the soil its ability to create a desirable chemical environment for crop growth and in some cases (not often, and certainly not generally) they may affect the physical character of soils. Soil fertility is crucially dependent on the availability of nutrients and in natural systems, plants growing in the soil subsist on the products of microbial activity; this is because the microorganisms are continuously oxidising the dead plant remains and leaving behind (in a form available to the plant) the nitrogenous and mineral compounds needed by plants for their growth.

One of the most important functions a soil performs for a growing crop is catching water during periods of rainfall and storing it for plants to use at a later time. If a soil is not capable of performing this task it is of limited value for non-irrigated crop production.

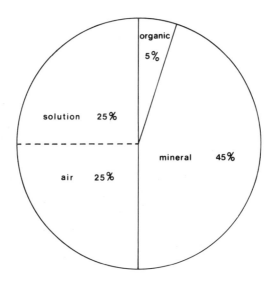

Fig. 1.3 Volume composition of an agricultural soil (15-20 cm in depth), representing favourable conditions for plant growth. The fractions occupied by air and water are extremely variable and their proportion largely determines its suitability for plant growth.

1.5.1 Soil texture and structure

Soil texture refers to the relative proportions of the sand, silt and clay (soil separates) in the soil. Extremes of texture, where one of these particle size separates is dominant, usually gives rise to poor moisture relationships and fertility. Sandy soils cannot hold sufficient water or nutrients, whereas fine clays may hold water and nutrients so tightly that they are unavailable for plant growth. Intermediate textured soils, called loams, provide high agricultural productivity.

Soil structure refers to the organisation of individual particles into clumps or aggregates. This is desirable agriculturally and important because it affects the soil's permeability and aeration capacity. Well-structured soils are more permeable than other soils of the same texture but lacking structure, because the larger pores *around* the aggregates conduct water rapidly, while small pores *within* aggregates hold water against drainage, so that it can be used by plants.

Structure is promoted by physical factors such as: freezing and thawing; wetting and drying; and by chemical compounds which tend to cement soil particles together, e.g. some organic products of microbial metabolism, and aluminium and iron hydroxides. Aggregate formation is greatly encouraged by the accumulation and decay of organic matter. This is partly due to its physico-chemical character, but also because its presence promotes an active system of soil organisms so essential to the soil's fertility and productivity.

1.5.2 Soil organic matter

Soil organic matter may be defined as any living or dead plant or animal materials in the soil and represents about 3-5% by weight of a typical mineral topsoil. It is a term often confused with humus which refers to the more or less stable decay products arising from the decomposition of the organic matter. Humus is usually black or brown in colour and is colloidal in nature.

Because organic matter stems from living organisms, it follows that it must contain all the nutrients needed for the growth of these organisms. Hence its composition directly reflects its initial source: it represents a major source of phosphorus and sulphur and is essentially the sole natural source of nitrogen; it also contains all the remaining elements essential to plant growth. However, unless the organic matter is decomposed, these nutrients are not available; the decomposition is carried out by soil organisms.

One of the key features of organic matter is its role in nutrient cycling. Nutrients are taken up by the plant roots and become incorporated in the organic matter of the plant. When the plant dies its remains fall to the ground and decompose, while roots decompose *in situ*. Nutrients are released into the soil by decomposition and pass from soil to plant and plant to soil in an endless cycle. When leaching occurs this represents a loss from the cycle though this may be offset by gains from soil minerals and atmospheric inputs. However, if the rate of loss exceeds the rate of gain then the store of nutrients within the cycle will slowly become depleted. The cycling process is particularly important for nitrogen, because of its absence from soil minerals.

Organic matter also functions as a 'granulator' of the mineral particles; this enhances the structure of the soil and leads to the loose friable condition associated with productive soils.

1.5.3 Soil chemical properties

Most of the chemical properties of soils are associated with a relatively small proportion of the total soil components — the clay and the humus, which are colloidal substances with a very high surface area. These display an important characteristic known as the cation-exchange capacity. Each of these materials has sites with negative charges which attract positively charged ions (cations). Such cations are called exchangeable if they can be replaced with other cations dissolved in the soil water surrounding the particles. Such replacement is possible if the bonding is not too strong and if the sites are accessible to the solution. Equilibria existing between the cation-exchange sites and the soil solution allow the concentrations of adsorbed cations to largely determine the concentrations of ions held in the soil solution; thus it influences the supply of nutrients to plant roots.

The capacity to retain and exchange cations in solution is an extremely important property affecting the soil's fertility (or potential fertility) and its ability to assimilate waste materials (section 4.4). The capacity is measurable, i.e. the cation-exchange capacity, and is usually expressed in terms of milliequivalents (of cations) per 100 g of soil (usually abbreviated as meq/100 g).

Representative cation-exchange capacities of clay-minerals and humus are given in table 1.2. This information makes it possible to roughly estimate the cation-exchange capacity of a soil of known composition. For example, a soil containing 4% organic matter and 20% clay (of which 75% is montmorillonite and 25% is illite) would have a cation-exchange capacity of 21.5 meq/100 g of soil calculated as follows:

4% organic matter:	0.04×200	$= 8.0$ meq
15% montmorillonite:	0.15×80	$= 12.0$ meq
5% illite:	0.05×30	$= 1.5$ meq
	total	21.5 meq

The composition is representative of many of the grassland soils of central United States (Thompson and Troeh, 1973).

Table 1.2   Cation-exchange capacities of soil clay and humus

	Cation-exchange capacities, meq/100 g	
	Representative	Usual range
Kaolinite	8	3 — 15
Chlorite	30	20 — 40
Illite	30	20 — 40
Montmorillonite	80	60 — 100
Allophane	100	50 — 200
Vermiculite	150	100 — 200
Humus	200	100 — 300

Inspection of table 1.2 shows that humus colloids have a greater cation-exchange capacity — and hence a greater nutrient-holding capacity — than many clays. However, there is generally more clay than humus in the soil; on balance their contribution to the chemical and physical properties of the soil are approximately equal. Some very sandy soils have virtually no cation-exchange capacity and therefore nutrients are easily leached because of the low retention.

The numerical value of the cation-exchange capacity is not necessarily a guide to a soil's potential concentration of valuable plant nutrients; this is because some of the exchange sites are occupied by H^+, Al^{3+} or some other non-essential ions; the remainder are usually dominated by the nutrient cations Ca^{2+} and Mg^{2+}.

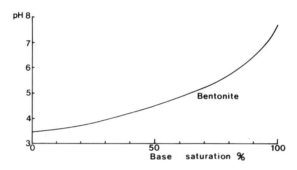

Fig. 1.4 pH-base saturation curve for montmorillonite (bentonite)

The percentage base saturation of a soil is usually taken as a good measure of how much of the cation-exchange capacity is being utilised to store plant nutrients. It is defined as the percentage of the cation-exchange capacity occupied by basic cations, usually Ca^{2+}, Mg^{2+}, K^+ and Na^+; it specifically excludes H^+ and Al^{3+} because they produce acid reactions. The percentage base saturation increases with pH (defined in section 1.5.4) as shown in fig. 1.4. In addition the cation-exchange capacity also increases with the pH of the soil.

Another important characteristic of the cation-exchange capacity is that it acts as the principle buffer mechanism in the soil. For example, pH is a measure of the portion of H^+ ions in the soil solution as distinct from those held on the micelles (minute silicate-clay colloid particles, generally carrying a negative charge). Also most soils have over 100 times as many cations adsorbed on the micelles as in true solution and there exists an equilibrium between these phases. If an acid is added to soil, only a fraction of the added H^+ ions remain in solution, the remainder displacing adsorbed bases on the micelles. The reverse happens when a base is added to the soil; bases displace H^+ from the micelles and these combine with OH^- ions to form water. Thus when an acid or base is added to the soil there is only a slight change in the portion of H^+ ions in the soil solution, and hence in the pH value. This characteristic is important when considering adding chemicals, such as lime, to adjust the soil pH; an estimate of the lime requirement can be based on trends as shown in fig. 1.4 together with a knowledge of the soil cation-exchange capacity (see Thompson and Troeh, 1973).

1.5.4 Soil acidity and pH

The condition of the soil measured in terms of its acidity, neutrality or alkalinity is one of its most important properties.

Acid or 'sour' soils are common in all regions where precipitation is high enough to leach appreciable amounts of exchangeable bases from the surface layers of the soil.

Alkalinity arises because of the high degree of base saturation; such soils are characteristic of most arid and semi-arid regions. Alkalinity also arises when the parent material of the soil is high in calcium, e.g. in limestone soils.

The reason soils are acid is because of the excess of H^+ ions in the soil solution. The H^+ ion activity is generally stated in terms of pH values which are defined as pH = $\log (1/[H^+])$ where $[H^+]$ is the hydrogen ion concentration. The pH is closely related to the amount of acidic cations (H^+ and Al^{3+}) and bases on its cation-exchange sites. Hydrogen ions may be liberated into the soil solution in a great variety of ways. For example, exchangeable Al^{3+} ions held on the micelles can be displaced by other cations and hydrolyse in the soil solution. Reactions are of the form:

$$Al^{3+} + H_2O \rightarrow Al(OH)^{2+} + H^+$$

In some derelict land materials containing iron sulphide (pyrite), there may be very considerable acidity with the pH between 2 and 4 which, when it weathers, produces sulphuric acid. Fertilisers containing sulphur and nitrogen also tend to lower the soil pH. Nevertheless, the pH can be raised by an application of lime if it is too low.

Vegetation itself influences soil pH because it provides decomposable organic matter which produces acids, and because it influences leaching. However, leaching is partially offset by the plants' ability to absorb nutrients from the root zone and to redeposit these at the soil surface when the plant dies, or in the form of litter. In general, grasses use more bases than trees and therefore deposit more bases on the soil surface than trees. Thus grasses help to keep the soil from becoming strongly acid. Many coniferous trees are light feeders on bases and tend to produce acid soils.

Most soils have pH values between 4 and 9 as shown in fig. 1.5.

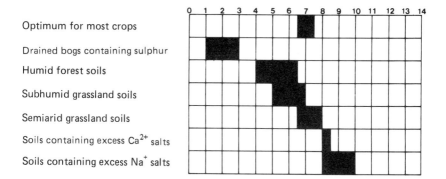

Fig. 1.5 Typical pH ranges for various types of soils

1.6 Plant growth and nutrition

The nutrients essential for plant growth are listed in table 1.3. All the elements except carbon and oxygen are obtained by terrestrial plants from the soil system and enter the plants through the roots. The macronutrients are required in large quantities; others, termed the micronutrients, are required only in trace amounts. The remaining elements shown in table 1.3 may be essential for particular species. It is difficult to be precise about the levels of available nutrients in the soil which are considered satisfactory for growth of common species, though table 1.4 shows examples of problem levels in unproductive soils.

Lack of any essential elements causes the plants to grow poorly and show deficiency symptoms. High concentrations of nutrients and other elements such as the heavy metals (section 4.9) may be toxic and cause growth to decrease. Figure 1.6 shows an idealised curve relating nutrient supply and growth. For some plants the optimum range of adequate nutrient concentration e.g. boron, is quite narrow. In practice the scheme depicted in fig. 1.6 is inadequate because of the interaction between nutrients. Nevertheless, it may be considered that the level of crop production can be no greater than that allowed by the most limiting of the essential plant growth factors (see Russell, 1973).

Table 1.3   Elements essential to or related to plant growth

	Element symbols	*Comments*
Macronutrients		
1. Carbon	C	
2. Hydrogen	H	Supplied by air and water
3. Oxygen	O	
4. Nitrogen	N	Supplied by atmosphere (through nitrogen fixation) and fertilisers
5. Phosphorus	P	
6. Potassium	K	
7. Magnesium	Mg	Supplied by soil minerals and fertilisers
8. Calcium	Ca	
9. Sulphur	S	
Micronutrients		
1. Iron	Fe	
2. Manganese	Mn	
3. Boron	B	Supplied by soil minerals and fertilisers;
4. Molybdenum	Mo	can cause crop failures if absent; can
5. Copper	Cu	become toxic to crops if present in large
6. Zinc	Zn	quantities
7. Chlorine	Cl	
Miscellaneous elements		
1. Sodium	Na	May substitute for potassium
2. Silica	Si	May accumulate in some plants
3. Cobalt	Co	Seldom needed except by some legumes
4. Aluminium	Al	Have never been shown to be essential for
5. Iodine	I	plant growth and can be harmful if present in large quantities

1.6.1 Nutrient supply and availability

Nutrients available to the plant must be present in ionic form, either in solution or held on the surface or between layers of soil colloids; the latter are obtained by an exchange process of the type described in sections 1.5.3 and 4.4.4. The positively charged ions, such as K^+, are mostly adsorbed and only a small fraction of the cations are found in the soil solution. Nitrate is the only negatively charged ion that wholly occurs in the soil solution; most of them, like the cations, are strongly adsorbed, mainly through forming insoluble compounds

Table 1.4 A basis for comparison: physical and chemical characteristics of some British and world soils

	Particle fraction			Loss on ignition (%)	pH	Cation-exchange capacity (meq/100 g)	Available plant nutrients (p.p.m.)							Mineralisable N (p.p.m.)	Total N (%)
	sand (%)	silt (%)	clay (%)				K	Ca	Mg	Fe	P	NH₄-N	NO₃-N		
British soils: productive															
Silty loam (Lincs)	40	35	25	3	6.5	25	350	2000	300	10	30	15	10	300	0.2
Sandy loam (Cheshire)	65	20	15	2.5	5.5	14	100	700	80	40	10	10	5	140	0.1
Clay loam (Cambs)	30	40	30	2.5	6.0	28	300	2500	600	60	15	20	5	[150]	0.2
Alluvial meadow (Yorks)	35	40	25	7	5.5	30	250	3000	350	5	25	20	3	100	0.4
unproductive															
Chalk grassland (Berks)	42	28	30	7	8.0	20	120	>20000	600	2	14	<1	20	60	0.3
Upland pasture (Dyfed)	50	35	15	8	5.0	15	100	250	100	50	3	25	10	120	0.2
Nardus fell (Cumbria)	55	15	30	12	4.5	[25]	80	160	50	220	2	30	1	20	0.2
Sandy heath (Norfolk)	75	15	10	6	4.5	[10]	60	250	40	60	3	8	2	[50]	0.2
Worldwide soils: productive															
Prairie silt loam (U.S.)	[20]	[65]	[15]	4	6.5	26	250	2000	500	100	50	[5]	[10]	[150]	0.25
Steppe chernozem (U.S.S.R.)	20	55	25	7	7.2	24	400	3000	300	30	30	[5]	[10]	[100]	0.2
unproductive															
Rain forest podsol (Borneo)	[20]	[10]	[70]	12	5.0	28	200	150	100	50	6	30	5	[50]	0.1
Savanna latosol (W. Africa)	[60]	[30]	[10]	4	6.0	8	100	800	250	250	1	[4]	[2]	[10]	0.07
Moraine (Signy, Antarctica)	30	10	60	1	6.8	6	40	600	80	50	5	1	<1	[5]	0.02
Levels for normal plant growth low				5		10	100	500	50	5	5	2	2	50	0.1
high				7.5		30	300	2000	300	200	20	20	20	200	1.0

Values underlined are those likely to be the cause of low productivity in common species

Values in brackets are estimates from other similar sites

Analytical methods are those given in Allen *et al.* (1974)

p.p.m. refers to mass concentration, i.e. mg/kg oven dry soil

with Al, Fe, Cu rather than by the exchange process. The retention of sulphate reflects both mechanisms.

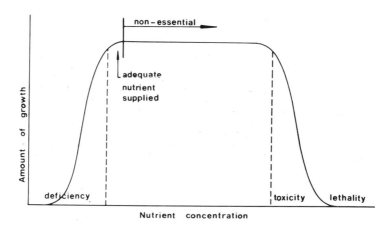

Fig. 1.6 Idealised relationship between nutrient concentration and the growth of the plant

Only a minute fraction of the elements are present in available form, and exist as complex and rather insoluble compounds. Availability depends on their transformation into simpler and more soluble forms. Unfortunately these are also susceptible to leaching and tend to disappear in drainage in humid areas.

Of the macronutrients, practically all of the nitrogen and much of the sulphur and phosphorus are held in organic combinations, whereas most of the potassium, calcium and magnesium exists in the soil in strictly inorganic forms. Nutrients held in organic forms become available as a result of their decomposition by soil microorganisms. Inorganic nutrients are generally released through weathering.

1.6.2 The nutritional importance of soil pH

Except in extreme cases, soil pH influences nutrient absorption almost wholly through its influence on nutrient availability and the presence of toxic forms. Several essential elements tend to become less available as the pH is raised from 5.0 to 7.5 or 8. Iron, manganese and zinc are good examples. At pH values below 5.0 to 5.5, aluminium, iron and manganese are often soluble and are present in sufficient quantities to be toxic to the growth of some plants. Figure 1.7 indicates that there is also likely to be a calcium deficiency in very acid soils. Phosphorus in the soil is never readily soluble but appears to be most available in a pH of around 6.5. Russell (1973) comments that if pH is suitably adjusted for phosphorus, other plant nutrients, if present in adequate amounts, will be satisfactorily available in most cases.

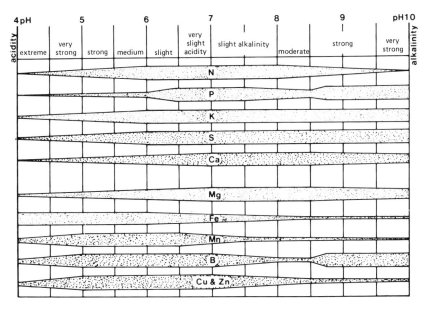

Fig. 1.7 General relationship between soil pH and nutrient availability. The width of the band at a particular pH value indicates the relative favourableness of this pH value (and associated factors) to the presence of the elements in question being in readily available forms (the wider the band the more favourable the circumstances); the band width does not necessarily reflect the actual amount present, this being influenced by other factors, such as cropping and fertilisation.

1.7 Ecological factors

Whether planning the restoration of disturbed land or planting for erosion control, it is vital that the selection of species and site pre-paration are consistent with short-term and long-term ecological constraints. Ecology is concerned with the relationship of living systems to their environment. Plants, like other organisms, thrive in an environment with fairly definable limits — limits such as those illustrated in fig. 1.6; species vary in their range of tolerances. Some important components of the environment are moisture, temperature, light, soil and living organisms. All components operate separately and jointly to influence plant distribution and behaviour, and they may be considered in terms of plant communities, zonation and succession.

1.7.1 Plant communities

Some species are so vigorous and successful that they oust direct competitors and dominate the vegetation over wide areas where they find the conditions of life favourable. Nevertheless in any piece of vegetation dominated by a single kind there are practically always other kinds associated with the dominant one; the associated plants find their life requirements satisfied by their situation. Overall, the whole collection (associated plants together with the dominants) form a plant community. The dominant plants of a natural community are generally the largest and tallest plants which can establish themselves and flourish in the particular habitat; this is because they overshadow the rest of the vegetation, excluding some species by the reduction of the light, and favouring the growth of other species which require shade (Tansley, 1968).

1.7.2 Zonation and succession

Zonation refers to the spatial extent of species and may be permanent (static) or in a state of flux (dynamic). Though zonation is apparent, it should be stressed that plant communities are not always perfectly defined entities with sharply defined boundaries, but show every grade of variation.

The natural communities associated with static zonation represent climax communities and consist of vegetation which is in equilibrium with its environment. They are stable forms of vegetation and are closely fitted to the climates in which they prevail. If the climate of a region changes, the vegetation changes with it. In many parts of the world the climatic climax is forest and this may represent the ultimate goal for stabilising, say, sand-dunes (Chapman, 1949).

Developmental or dynamic zonation is commonly called plant succession. This arises when the entrance and survival of a species in a new habitat changes the nature of the habitat, thus permitting the entry of other species until a climax stage is reached. To illustrate these phenomena it is useful to consider the pattern of vegetation developing on bare land or from water, because these provide insight into the strategy for revegetating a disturbed site or the banks of a river.

When a piece of bare ground is exposed, the colonisation often begins with lower plants — algae, lichens and mosses, perhaps, on damp rock surfaces together with annual flowering plants which usually figure as pioneer colonists on loose soil. The decay of the bodies of the first plants to settle contributes humus to the raw mineral soil and begins to improve it. With time, perennial plants take root and gradually establish a closed carpet of vegetation. Among the perennials, the seedlings of woody plants such as briars and brambles are found; where these take root and flourish, they begin to over-shadow the lower-growing plants and kill those requiring full illumina-tion. With further soil development, climate and other factors per-mitting, trees may grow to form the final climax. In this series the soil becomes damper owing to the water-holding power of the accumulated humus (Tansley, 1968).

A different succession begins in water whereby the early stages of colonisation are by water plants. When the ground level is raised by accumulation of plant debris or by silting, reeds and sedges become established, forming a reed-swamp. With a further rise of the soil, marsh plants settle amongst the plants of the reed-swamp and gradually supersede them. Above the water level, shrubs such as sallows and alders grow in saturated or near-saturated soil forming marsh wood-land. With further accumulation of humus or silting from flood water the surface levels become drier and suitable for the invasion of other trees such as birch, ash and oak, so that eventually a forest may be established. In this progression, the soil becomes drier up to a certain point and the succession beginning in water through its various stages to the climax is called a hydrosere. The series beginning on a dry habitat and progressing to a climax community is known as a xero-sere. Thus the two types of succession converge and the climax community may be practically the same in both cases (Tansley, 1968).

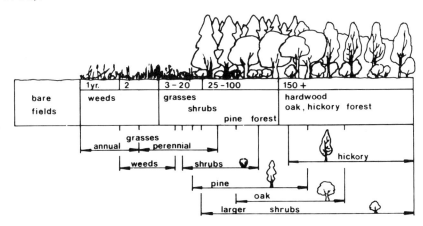

Fig. 1.8 Typical plant succession

A natural succession may take up to and beyond a century to reach the climax state (fig. 1.8) and there is clearly an incentive to foreshorten this timescale in a land restoration project! This subject is pursued in chapter 3.

1.8 References

Allen, S.E., Grimshaw, M., Parkinson, J.A. and Quarmby, C. (1974) *Chemical Analysis of Ecological Materials*. Oxford: Blackwell.

Bradshaw, A.D. and Chadwick, M.J. (1980) *The Restoration of Land*. Oxford: Blackwell.

Buckman, H.O. and Brady, N.C. (1969) *The Nature and Properties of Soils*. 7th edn. London: Macmillan.

Chapman, V.J. (1949) 'The stabilisation of sand-dunes by vegetation'. *Proceedings of the Conference on Biology in Civil Engineering*. London: Institution of Civil Engineers, 142-57.

Etherington, J.R. (1975) *Environment and Plant Ecology*. New York: Wiley.

Hansen, V.E., Israelsen, O.W. and Stringham, G.E. (1980) *Irrigation Principles and Practices*. New York: Wiley.

Kramer, P.J. (1969) *Plant and Soil Water Relationships: A Modern Synthesis*. New York: McGraw-Hill.

Marshall, T.J. and Holmes, J.W. (1979) *Soil Physics*. Cambridge: Cambridge University Press.

Mehlich, A. (1941) 'Base unsaturation and pH in relation to soil type'. *Soil Science Society of America, Proceedings* 6, 150-56.

Meidner, H. and Sheriff, H.W. (1976) *Water and Plants*. Glasgow: Blackie.

Russell, E.W. (1973) *Soil Conditions and Plant Growth*. 10th edn. London: Longman.

Sopher, C.D. and Baird, J.V. (1978) *Soils and Soil Management*. Virginia: Reston.

Tansley, A.G. (1968) *Britain's Green Mantle: Past, Present and Future*. 2nd edn. London: Allen and Unwin.

Thompson, L.M. and Troeh, F.R. (1973) *Soils and Soil Fertility*. 3rd edn. New York: McGraw-Hill.

Truog, E. (1946) 'Soil reaction influence on availability of plant nutrients'. *Soil Science Society of America, Proceedings* 11, 305-08.

van't Woudt, B.D. and Hagan, R.M. (1957) 'Crop responses at excessively high soil moisture levels'. Luthin, J.N. *ed. Drainage of Agricultural Lands. Agronomy* 7, 514-611. Madison, Wisconsin: American Society of Agronomy.

Weier, T.E., Stocking, C.R. and Barbour, M.G. (1974) *Botany: An Introduction to Plant Biology*. 5th edn. New York: Wiley.

Winter, E.J. (1974) *Water, Soil and the Plant*. London: Macmillan.

Young, W.C. (1968) 'Ecology of roadside treatment'. *Journal of Soil and Water Conservation* 23(2), 47-50.

2 Vegetation and hydrology

2.1 Introduction

Water is a unique resource central to many problems in the planning of land use. Throughout history human populations have occupied land and converted vegetation and soils to their own uses without sufficient regard for the hydrological consequences. Often this has resulted in flooding, droughts, soil erosion and salinity, bringing disaster to the communities causing them. Man's transition from a nomadic hunter to a settled existence based on agriculture has been accompanied by a continuous change of land use. Forests have been cut, burned and cleared; natural grasslands and brushlands have been grazed, burned and ploughed. Over vast areas natural vegetation have been replaced with cultivated plants and tillage and this has created a new kind of soil, periodically bare-surfaced; roads have been laid as impervious strips, and towns and cities continue to expand; water has been diverted from stream channels; meadows and valley basins have been drained of their stored water. All these and other activities of man have contributed to changes in the quantity, regime and quality of water yielded by the land. Where these changes are skilfully done, the water resources may not be damaged, and may even be enhanced. Unfortunately very few of the watersheds cleared of their natural vegetation have been skilfully managed. Particularly tragic has been the destruction of large tracts of ancient forest.

Forest vegetation is usually the dominant vegetation in high rainfall areas, which are also the prime source of river flow. To destroy natural forest is to destroy nature's first line of defence against the ravages of water flow following heavy rainfall. This is amply illustrated by Hicks' (1975) account of the disastrous flooding of the River Arno which hit Florence in 1966, when 300 persons lost their lives:

The hills of Tuscany in Etruscan times were densely forested, and the Arno ran clear the whole year round. The growth of the city led to greater demands for timber, particularly in Christian times, and to grazing of the hills by goats and by sheep, the latter for the Florentine and wool trade. Eventually the denuded, sunbaked hills no longer retained moisture and the Arno dried up during the long hot summers, but deluged the town with mud from the eroded hills during floods. Records show that a minor flood occurred every twenty-four years, and that a major disaster occurred roughly every century. After the 1333 disaster (amongst the earliest of the major floods), a certain Vico del Cilento recommended reafforestation of the hills. The city government of Florence however demurred.

These same bare hills, their slopes scarred by landslides, still preside over the fate of Florence. The mathematics of probability do not exclude a repetition of the 1966 disaster at any time. . .

Changes in land use inevitably disturb the natural balances controlling water movement within the hydrologic cycle and all have pronounced and well-documented effects on the hydrologic system. Decreased infiltration, increased run-off and accelerated erosion are some of the more obvious effects. Less apparent, but no less important are changes in the pattern of evaporative transfer – the mechanism solely responsible for the replenishment of atmospheric water lost by precipitation. Whether dealing with flood control, irrigation management or catchment water yield, it is vital to evaluate the evaporative demand and its controls.

In most situations of changing land use, one is not so much concerned with changes in individual processes, as in their overall effects through space and time; this is most easily seen by considering the hydrologic cycle and surface water balance.

2.1.1 Hydrologic cycle
The hydrologic cycle, shown schematically in fig. 2.1, describes the ways in which water moves around the earth. During a continuous circulation driven by solar energy, water evaporates from land and water surfaces, condenses, and is returned as precipitation – generally as rain or snow. Over the oceans, evaporation and precipitation almost balance, with a small but significant fraction contributing to the precipitation over land. Of the global precipitation on land approximately 2/3 is lost by evaporation and 1/3 by runoff. While averaged data of this nature has little practical significance in applied hydrology, it identifies the interdependence of two key processes

and emphasises the importance of evaporation in surface hydrology. Since much of the land surface is covered by vegetation, it is reasonable to assume that this has a significant impact on the water transfer. To understand its potential influence, it is necessary to examine the surface water balance, i.e. the inputs and outputs in greater detail. Discussion will focus on rain rather than snow, since once it has melted, the snowmelt follows the same pathways as rainwater.

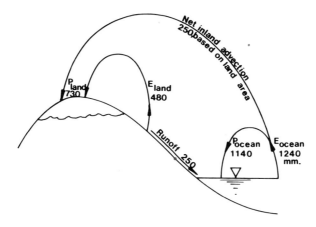

Fig. 2.1 Hydrologic cycle: Annual average of fluxes in mm equivalent depth (based on data from Lvovitch, 1972). Symbols: P = precipitation; and E = evaporative flux

2.1.2 Surface water balance

Before reaching the earth's surface much of the rain is intercepted by vegetation. Some of the water is trapped due to surface wetting and is subsequently evaporated before reaching the ground – a process which is known as interception. The excess seeps into the ground or fills small depressions which may eventually spill – leading to overland flow. Both infiltration, i.e. ground absorption, and overland flow are influenced by the vegetative cover. Within the ground, water held by capillary forces is subsequently evaporated from the soil surface; of greater significance is its withdrawal by root suction and loss by transpiration. Water which cannot be held by capillary forces percolates into the soil, to re-emerge as subsurface storm runoff (interflow), or enters the groundwater zone where the pores are completely filled with water.

The combination of these processes can be represented by the water balance:

$$P_g = I + E + OF + \Delta SM + \Delta GWS + GWR \qquad [2.1]$$

where the symbols (left to right) expressed as equivalent depths of water over some time interval represent gross precipitation, interception, evaporation (excluding interception), overland flow, change in soil moisture, change in groundwater storage and groundwater runoff. Equation [2.1] is applicable to a small drainage basin underlain with an impervious layer at depth as depicted in fig. 2.2.

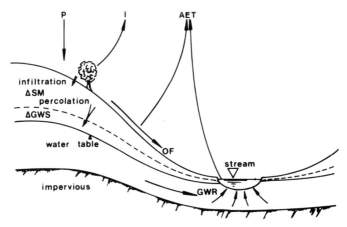

Fig. 2.2 Components of the water balance on a slope underlain by an impervious layer. Symbols: P = precipitation; I = interception; AET = actual evapotranspiration; OF = overland flow; ΔSM = change in soil moisture; ΔGWS = change in groundwater storage; GWR = groundwater runoff.

Over a lengthy period of time the terms ΔSM and ΔGWS fluctuate about a mean value close to zero and are negligible; this yields the approximation:

$$P_g - E^* = R \qquad\qquad [2.2]$$

where E^* is the total evaporation $(I + E)$ and R is the runoff (comprising OF and GWR). It will be seen later that it is difficult to consider the terms I and E separately. Overall equation [2.2] emphasises the important interdependence of evaporation and runoff which was evident in fig. 2.1.

In this chapter, initial discussion will focus on the components of equation [2.1], highlighting the particular influences of vegetation. This is followed by a brief discussion of hydrology and land use, casting a general perspective on a number of situations encountered in the agricultural domain and in forest lands. Though primarily intended for readers with a grounding in basic hydrology, the concepts discussed are fairly straightforward and can be amplified by

referral to the quoted references, or to texts such as Ward (1975) and
Dunne and Leopold (1978).

2.2 Evaporation

Inspection of fig. 2.1 indicates that evaporation is the dominant loss
mechanism for water over land surfaces, and on a global scale the
average precipitation exactly balances average evaporation. Evaporation of water represents a change of state from liquid to vapour; for
this to occur there must be a supply of heat of vaporisation, λ (=
2.477×10^6 J kg^{-1} at 10°C). Hence the overriding condition for
maintained evaporation is the provision of a source of energy. Shor-
tening a discussion which may be obtained in full elsewhere, e.g.
Dunne and Leopold (1978), the source of energy is solar — varying
with season and latitude according to site; and with cloudiness, slope,
aspect and topography.

2.2.1 Radiant energy balance

The radiant energy balance for a given site, in its basic essentials, can
be written verbally as:

Balance	Gains		Losses
Net radiation =	$\begin{bmatrix} \text{Incident short-} \\ \text{wave radiation} \end{bmatrix}$	$-$	$\begin{bmatrix} \text{Reflected short-wave} \\ \text{radiation} \\ + \\ \text{Emitted long-wave} \\ \text{radiation} \end{bmatrix}$

The incident short-wave radiation Q_s consists of direct solar and
diffuse radiation from the sun. If the albedo (the reflection coefficient) of the surface is α, the reflected short-wave flux is αQ_s. The
emitted long-wave radiation Q_{lw} arises because all bodies possessing
energy, i.e. with temperatures above absolute zero (-273.2°C), must
emit radiation. With these inclusions the radiation balance may be
written:

$$Q_n = Q_s (1 - \alpha) - Q_{lw} \tag{2.3}$$

where Q_n is the net radiation, representing the flux of energy available for evaporation and sensible heat transfer, i.e. heat energy which
causes a change in temperature and not a change of state.

Inspection of table 2.1 shows that the albedo depends on the type

of surface. In the case of fresh snow, much of the incident short-
wave radiation is reflected, which contrasts with a water surface
where much of the energy is absorbed. Forest vegetation generally
possesses a lower albedo than grass and agricultural crops (typically
0.25 for green vegetation of this type). This implies that there is
likely to be greater input of radiation to forested areas of a catchment compared to the input to areas covered by grass. If the long-
wave losses are treated as equal for both vegetation types then
elimination of Q_{lw} from equation 2.3 leads to:

$$Q_n^{forest} = Q_s^{grass} + (\alpha^{grass} - \alpha^{forest}) Q_s^{grass} \qquad [2.4]$$

This illustrates that changes in the pattern of vegetation over an area
alter the input of net radiation.

Table 2.1   Albedo of different surface covers (based on data from Baumgartner
(1972) and Oke (1978))

Surface	Albedo
Snow: old	0.40
fresh	0.85 – 0.95
Grass	0.17 – 0.30
Agricultural crops	0.15 – 0.28
Deciduous forests	0.10 – 0.18
Coniferous forests	0.04 – 0.15
Water surfaces	0.03 – 1.00
Bare soil	0.06 – 0.48

In equation [2.3] the net back-radiation can be expressed in terms
of air temperature, atmospheric humidity and cloudiness and can be
estimated from a number of empirical equations. The most common
of these is cited in Dunne and Leopold (1978) as equivalent to:

$$Q_{lw} = \sigma T_2^4 (0.56 - 0.08 \sqrt{e})(0.10 + 0.90\, n/N) \qquad [2.5]$$

where σ = the Stefan-Boltzmann constant (5.57×10^{-8} W m^{-2} K^{-4})
    T_2 = air temperature 2 m above the surface (K)
    e = vapour pressure 2 m above the surface (mbar)
    n = measured duration of bright sunshine (h)
    N = maximum possible duration of bright sunshine for latitude
      and time of year (h)

The term Q_s can be estimated by:

$$Q_s = I_0 \, (a + bn/N) \qquad [2.6]$$

where a, b = empirical constants (Glover and McCulloch (1958) cite
$\qquad a = 0.29$ cosine ($^\circ$ latitude) and $b = 0.52$)
$\quad I_0$ = solar radiation per day received on a horizontal
$\qquad$ surface at the exterior of the atmosphere (W m^{-2})

Values for n, N may be obtained from standard meteorological summaries, e.g. Ministry of Agriculture, Fisheries and Food (1976) and I_0 from List (1968). It should be noted that estimates from these empirical equations are not very precise and errors can exceed ± 25%. Q_n can be measured directly with a net radiometer and such data are published for a few important meteorological stations. A third method involves estimating net radiation as a linear function of Q_s for a given surface as exemplified by Szeicz et al. (1969). Q_s can also be measured directly by means of a pyrheliometer.

2.2.2 Energy balance of small pans and shallow lakes

Historically, transpiration losses arising from plant surfaces have been assessed in terms of the equivalent evaporation taking place from a free water surface under similar climatic conditions. The latter can be monitored directly by measuring the water losses from standard evaporation pans or calculated using an energy balance. For small bodies of water, such as within evaporation pans or in shallow lakes, Penman (1948) argued that the energy exchange could be approximated by:

$$Q_n = Q_h + Q_e \qquad [2.7]$$

where Q_h = sensible heat transfer, i.e. heat energy which causes a
$\qquad$ change of temperature and not a change of state
$\quad Q_e$ = energy used for evaporation, i.e. latent heat transfer

By dividing by $\rho\lambda$ in which ρ is the density of water, these energy components can be expressed in terms of equivalent depths as:

$$H = C + E_0 \qquad [2.8]$$

on the basis of which Penman showed that evaporation from an open water surface could be expressed as:

$$E_0 = \frac{H\Delta + \gamma E_a}{\Delta + \gamma} \qquad \text{(mm day}^{-1}\text{)} \qquad\qquad [2.9]$$

in which H is the evaporative equivalent of net radiation at the surface (mm day^{-1}), and E_a is an empirical aerodynamic or ventilation term defined by:

$$E_a = 0.26\,(e_s - e)\,(1 + u_2/100) \qquad \text{(mm day}^{-1}\text{)} \qquad [2.10]$$

The quantities e and e_s (mbar) refer to the actual and saturation values of the vapour pressure at 2 m above the surface; u_2 signifies the corresponding wind run in miles per day; γ is the psychrometer constant (= 0.66 mbar K^{-1} near sea level); and $\Delta = de_s/dT$, the slope of the saturation vapour pressure versus temperature curve for water at air temperature.

As an aid to computation, the numerator and the denominator of equation [2.9] can be divided by γ, with values of the dimensionless parameter Δ/γ listed in table 2.2.

Table 2.2   Value of Penman's dimensionless parameter Δ/γ for various temperatures (derived from data given by Monteith (1973); Tables A3 and A4) and refer to a pressure of 1000 mbar.

$T(^\circ C)$	Δ/γ	$T(^\circ C)$	Δ/γ
0	0.70	25	2.85
5	0.94	30	3.67
10	1.27	35	4.67
15	1.68	40	5.87
20	2.20	45	7.32

2.2.3 Evapotranspiration: Penman equation

When computing the water loss from a catchment, it is usually impossible to separate the process of transpiration from direct evaporation from the soil surface and open water surfaces. These processes are usually considered together under the title evapotranspiration. In section 1.4.4 it was indicated that transpiration is a function of the normal physiological activity of a plant in which stomata effectively control the transpiration rate, and evaporation accounts for losses of water from free surface water, snow, ice and directly from the soil.

Penman (1948) expressed the evapotranspiration from short green vegetation completely covering the ground and adequately supplied with water at its roots as:

$$E_T = fE_0 \qquad [2.11]$$

where the empirical factor f was found to be 0.8 in summer and 0.6 in winter, the expression holding to within $\pm$ 15% in all temperate climates.

Rijtema (1965) demonstrated that the magnitude of f and its seasonal behaviour are consistent with the difference in albedo between an open water surface and short green vegetation so that equation [2.11] is simply an approximation to:

$$E_T = (\Delta H + \gamma E_a)/(\Delta + \gamma) \qquad [2.12]$$

where H is the evaporation equivalent of the net radiant energy input to the vegetation itself. Values of H are seldom directly available, but can be deduced accurately enough from meteorological records using equation [2.3] with the appropriate albedo. E_T is often referred to as 'potential evaporation' or 'potential evapotranspiration' and provides good estimates of the amount of water transpired by short green vegetation subject to the conditions stated. It should also be emphasised that the 'Penman' estimate was designed for monthly values and should not be used for daily values unless accurate radiation data (as distinct from sunshine data) are available. Approximations to the full calculation, based on a sunshine calculation are described in Ministry of Agriculture, Fisheries and Food (1967). Potential evaporation is not affected to any significant extent by the rate of growth and there is evidence that a wide range of plant types transpire at the same rate per unit ground area. However, there is mounting evidence, e.g. Thom and Oliver (1977) and Clark and Newson (1978), which indicates that the Penman estimate can produce serious errors when applied to tall vegetation. The evaporation rate is independent of the type of soil, but the length of time during which E_T can be maintained without soil moisture replenishment is certainly a function of soil type and soil management. If water is not freely available, actual evaporation will be less than E_T and the character of individual plant and soil types becomes more evident.

Estimates of E_T are used in several ways. For hydrological purposes evaporation dominates the water balance and it is necessary to assess the amount of water transpired or evaporated from the ground in order to assess factors such as soil moisture content, groundwater recharge and stream flow; the calculation of irrigation water requirements is based largely on estimates of E_T, both in the planning phase of a project and in the day-to-day control of water supply at the farm level (see section 2.7.1.3).

Example 2.1 Penman calculation of potential evaporation

Calculate the potential evaporation using the following data for the month of July at latitude 52°N: $T = 16.5°C$; $u_2 = 5.0$ m.p.h.; $e_2 = 14.67$ mbar; $n = 6$ h day^{-1}; $\alpha = 0.25$.

From standard tables, e.g. List (1968), e_s (at 16.5°C) = 18.76 mbar; $N = 16.1$ h; $I_0 = 458$ W m^{-2}.

From equation [2.6]:
$$Q_s = 458 (0.29 \cos 52 + 0.52 \times 6.0/16.1)$$
$$= 170 \text{ W m}^{-2}$$

From equation [2.5]:
$$Q_{lw} = (5.57 \times 10^{-8}) (289.5)^4 (0.56 - 0.08 \sqrt{14.67}) \times (0.1 + 0.9 \times 6.0/16.1)$$
$$= 43 \text{ W m}^{-2}$$

From equation [2.4]:
$$Q_n = 170 (1 - 0.25) - 43$$
$$= 84.5 \text{ W m}^{-2}$$

Hence,
$$H = \frac{Q_n}{\rho \lambda}$$
$$= \frac{84.5 \text{ J s}^{-1}\text{m}^{-2}}{(1000 \text{ kg m}^{-3}) (2.45 \times 10^6 \text{ J kg}^{-1})} \times (86\,400 \text{ s day}^{-1}) (1000 \text{ mm m}^{-1})$$
$$= 2.98 \text{ mm day}^{-1}$$

From equation [2.10]:
$$E_a = 0.26 (1 + 120/100) (18.76 - 14.67)$$
$$= 2.34 \text{ mm day}^{-1}$$
From Monteith (1973):
$$\Delta/\gamma \text{ (at 16.5°C)} = 1.83$$

From equation [2.12]:
$$E_T = \frac{(\Delta/\gamma) H + E_a}{(\Delta/\gamma) + 1}$$

$$= \frac{(1.83)\,(2.98) + 2.34}{1.83 + 1}$$
$$= 2.75 \text{ mm day}^{-1}$$

2.2.4 Transpiration: Penman-Monteith equation

Inspection of equation [2.12] shows that it contains no factor to take account of surface features other than the dependence of the available energy on the surface albedo. Transpiration, however, depends on evaporation taking place within the plant tissues with a subsequent diffusion of water molecules to the free atmosphere via the stomatal pores. These are able to open and close, and in doing so, exert a control on the transpiration rate (see section 1.4.4). It has also been noted that Penman's equation has significant defects when applied to tall vegetation.

Monteith (1965) developed Penman's approach by introducing a physiological control – the so called 'surface resistance' – into Penman's equation and redefined the ventilation term to take account of the aerodynamic roughness of the evaporating surface. The resultant Penman-Monteith equation represents an important milestone in our understanding of the effects of plants in controlling transpiration losses, and lies at the heart of present-day understanding of evaporative transfer from plant surfaces. When soil heat flux is neglected the Penman-Monteith equation is:

$$\lambda E = \frac{\Delta Q_n + \rho c_p\,(e_s - e)/r_a}{\Delta + \gamma\,(1 + r_s/r_a)} \qquad (\text{W m}^{-2}) \qquad\qquad [2.13]$$

where E = vapour flux (kg m^{-2} s^{-1})

$\quad\quad Q_n$ = net radiation (W m^{-2})

$\quad\quad \rho$ = density of air (kg m^{-3})

$\quad\quad c_p$ = specific heat of air (J kg$^{-1}\,^\circ$C^{-1})

$\quad\quad r_s$ = surface resistance (s m^{-1})

$\quad\quad r_a$ = aerodynamic resistance (s m^{-1})

with the remaining parameters as defined.

The factor r_a expresses the aerodynamic resistance to the diffusion of water vapour from the surface to some arbitrary reference point above the surface. The quantity $1/r_a$ increases linearly with wind speed and increases with the aerodynamic roughness of the surface. In both cases, it implies that the rate of diffusion of water vapour is increased and enhances the evaporation rate. For herbaceous communities (0.1 to 1 m height) and at moderate windspeeds, $r_a \sim 50$ s m^{-1} and is usually an order of magnitude less for trees, because of their larger surface roughness (see table 5.1).

The term r_s may be interpreted as the 'bulk' physiological resistance (per unit ground area) and in dry conditions closely reflects the behaviour of the leaf stomata. If one examines a single leaf, the evaporative flux λE is maintained by a vapour pressure difference $e_s - e_0$, where e_s is the saturated vapour pressure within the substomatal cavity (see figs. 1.1 and 2.3) and e_0 is the vapour pressure just outside the leaf. Making use of the resistance analogue noted in section 5.7 and Appendix 1, the evaporation process can be represented by:

$$\lambda E = \frac{\rho c_p}{\gamma} \frac{(e_s - e_0)}{r_{st}} \qquad [2.14]$$

where r_{st} is stomatal resistance. Treating a plant stand as an extended large leaf, r_s represents the integrated effect of stomatal resistance associated with individual leaves and the terms r_s and r_{st} are virtually synonymous. When the stomata are fully open, r_{st} (or r_s) takes a minimum value, whereas when the stomata are closed, r_{st} (and r_s) $\rightarrow \infty$ and transpiration ceases. The size and behaviour of the effective value of r_s is the subject of much research, being dependent on the vegetation type, its condition, the soil moisture deficit, the vapour pressure deficit, the time of year, the amount of intercepted precipitation and many other factors.

When foliage is thoroughly wet, $e_0 = e_s$, implying that the 'force' driving the transfer of water molecules through the stomatal pores is zero and transpiration (as distinct from evaporation) ceases. Because evaporation can still take place from the wet foliage surface, it is equivalent to the statement $r_s = 0$ (since $r_s \propto (e_s - e_0)/\lambda E$). Otherwise the surface resistance retains a significant residual value. For short vegetation, r_s is often similar in size to r_a; whereas for dry trees it can be considerably greater, e.g. for pine trees Szeicz et al. (1969) quote r_s as in the range 100-150 s m^{-1}, indicating very effective stomatal control of transpiration by pine trees.

Figure 2.4 shows the evaporation rate calculated from the Penman-Monteith equation for different values of r_a and r_s under a set of meteorological conditions corresponding to a cool summer day. Typically the evaporation rates from grass and coniferous trees during dry conditions are similar — the evaporation from grass being slightly higher. But if the surfaces are wet, the evaporation rates are far from similar since evaporation rates from grass increase by a factor 1.4-1.8, whilst for trees the increase is by a factor of 5-15 (see

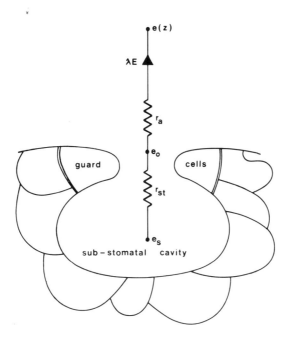

Fig. 2.3 Schematic cross-section through a portion of a leaf depicting the transfer of water vapour via stomatal pores. The equivalent latent heat flux must pass the stomatal resistance r_{st} (or r_s, the bulk surface resistance) as well as an aerodynamic resistance represented by r_a. Other symbols: e_s = saturation vapour pressure at the leaf temperature; $e(z)$ = saturation vapour pressure at distance z from surface; e_0 = saturation vapour pressure at the leaf surface

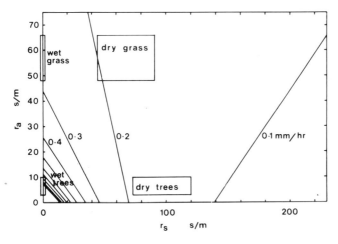

Fig. 2.4 Evaporation rates calculated from the Penman-Monteith equation [2.13] as a function of the aerodynamic resistance (r_a) and the surface resistance (r_s) for cool summer day-time conditions (net radiation = 200 W m^{-2}, vapour pressure deficit = 5 mbar; air temperature = 10°C)

section 2.2.7.2).

For practical purposes the Penman-Monteith equation, although theoretically capable of giving accurate estimates of transpiration from any crop, is seldom used. Usually this is not because of the scarcity of the necessary meteorological data (which are becoming increasingly available from automatic weather stations), but because of the lack of knowledge of the crop aerodynamic and surface resistances. At present, knowledge of the physiology of stomatal behaviour is insufficient to allow causal models to be constructed and recourse has to be made to empiricism. For example, Calder (1977) showed that a model in which the gross annual variation in r_s was cyclic and modified for changes in vapour pressure deficit, provided good agreement with observation for predictions of transpiration from a spruce forest.

Despite the present difficulties in using equation [2.13], it represents the foundation on which future developments are likely to be based. For example Thom and Oliver (1977) examined the discrepancies between the Penman and the Penman-Monteith equations and recommended an intermediate form which can be used for hydrological purposes.

Example 2.2 Penman-Monteith estimate of transpiration

Calculate the transpiration rate using the Penman-Monteith equation assuming the following data: net radiation $(Q_n) = 200$ W m^{-2}; vapour pressure deficit $(e_s - e) = 5$ mbar; air temperature = $10°C$; surface resistance $(r_s) = 50$ s m^{-1}; aerodynamic resistance $(r_a) = 50$ s m^{-1}; air density $(\rho) = 1.24$ kg m^{-3}; specific heat of air at constant pressure $(c_p) = 1020$ J kg^{-1} K^{-1}; psychrometer constant $(\gamma) = 0.65$ mbar $°C^{-1}$; latent heat of vaporisation $(\lambda) = 2.477 \times 10^6$ J kg^{-1} and change in saturation vapour pressure per $°C$ $(\Delta) = 0.83$ mbar $°C^{-1}$.

Substitution into equation [2.13] leads to:

$$\lambda E = \frac{0.83 \times 200 + 1.24 \times 1020 \times 5/50}{0.83 + 0.65 (1 + 50/50)}$$

$$= \frac{166 + 126.5}{2.13} \qquad = 137.3 \text{ W m}^{-2}$$

$$E = \frac{137.3}{\lambda} = \frac{(137.3 \text{ J s}^{-1} \text{ m}^{-2})(3600 \text{ s h}^{-1})}{(2.48 \times 10^6 \text{ J kg}^{-1})}$$

$= 0.2 \text{ kg m}^{-2} \text{ h}^{-1}$

$= 0.2 \text{ mm h}^{-1}$ (since $1 \text{ kg m}^{-2} = 1$ mm equivalent depth)

Note: this calculation is commensurate with the meteorological conditions cited in fig. 2.4. By altering the parameters r_s and r_a, the trends shown in fig. 2.4 may be generated.

2.2.5 *Soil water and its effect on transpiration*

In previous discussions it was noted that a plant adequately supplied with water transpires at the potential rate. But with the progressive depletion of soil moisture levels, plants find it increasingly difficult to withdraw water from the soil by root suction. To avoid dehydration, transpiration is restricted by reduced stomatal opening and allows the plant to reach a steady state in which root uptake and transpiration are equal. The limit of this automatic adjustment is set by wilting. Thus, 'actual' transpiration depends on a combination of plant and climatic factors, and the availability of water.

Whenever climatic conditions are such that potential evaporation exceeds the precipitation input, we should expect reductions in soil moisture levels; their consequent influence on the actual evaporation rate warrants careful attention in any problem involving predictions based on the surface water balance.

2.2.5.1 ROOT RANGE AND AVAILABLE WATER

In section 1.4 two important parameters were identified for characterising the water regime in the soil, namely:

(a) the field capacity, approximating to the retained water in near-saturated conditions (generally taken as a soil suction of 0.05 bar in the United Kingdom); and

(b) the permanent wilting point, corresponding to 15 bar suction.

The 'available' capacity, θ_a, is defined by the difference between these states:

$$\theta_a = \theta_v (0.05) - \theta_v (15) \qquad\qquad [2.15]$$

$\theta_v(\psi)$ denotes the volumetric moisture content of the soil corresponding to the moisture tension (suction) ψ and depends on the soil texture, as illustrated in table 2.3. Neglecting the extremes, $\theta_a = 0.20 \pm 0.05$ of the soil volume. Hence for a rooting depth of 1 m:

$$\frac{\text{available water}}{\text{capacity of root zone}} = \frac{\text{available water}}{\text{fraction}} \times \text{root depth}$$

$$= 0.20 \pm 0.05 \quad \times 1 \text{ m}$$

Thus 150-250 mm of water is available for transpiration. Given an evaporation rate of 3-5 mm day^{-1}, this reserve would be exhausted in 1-2 months.

Table 2.3   Mean values of field capacity, permanent wilting percentage and available water expressed as a fraction of the soil volume for different textural classes.

	Fraction of soil volume*		
	Field capacity	Permanent wilting	Available water capacity
Sand	0.09	0.02	0.07
Sandy loam	0.27	0.11	0.16
Loam	0.34	0.13	0.21
Silt loam	0.38	0.14	0.24
Clay loam	0.30	0.16	0.14
Clay	0.39	0.22	0.16
Peat	0.55	0.25	0.30

*Original data from Salter and Williams (1965) corresponds to top 0.3 m of soil

Table 2.4   Maximum soil moisture deficits for different types of vegetation

Crops	Maximum soil moisture deficit (mm)
Cereals, etc.	200
Root crops, etc.	150
Temporary grasses	100
Permanent grass	125
Rough grazing	50
Permanent woodland	250 but 125 on poor land

Moisture extraction patterns are more complicated than indicated by this simple model because of the root structure. When water is freely available, the highest rates of extraction coincide with the greatest concentration of roots and root hairs — these usually occurring in the upper layers of the soil. This means that if no water is added, the surface layers may reach wilting point while layers below are still above wilting point. Though the plant will still be able to gain water from the lower layers, it will not be supplied at the same high level as before and will not produce so well. It should also be recognised that the majority of valuable plant foods are in the surface layers; there would be a severe nutritional loss if these were not exploited by the plant.

The wilting point is sometimes expressed in terms of an equivalent soil moisture deficit (*SMD*). Generally, this refers to the equivalent depth of water required to restore the moisture level to field capacity. Experimental work has shown that for each type of crop there is a critical maximum soil moisture deficit corresponding to the available water capacity of the root zone. Examples of the maximum soil moisture deficit for various crops are given in table 2.4 and a more complete list is provided in Grindley (1969).

2.2.5.2 RESPONSE OF ACTUAL TRANSPIRATION

Experimental investigations of the way in which the transpiration rate responds to the drying of soil have led to a diversity of conclusions which Hillel (1971) summarised into the following categories, referring to fig. 2.5.

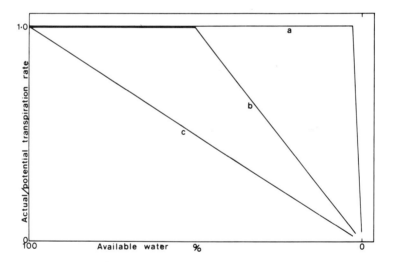

Fig. 2.5 Three hypotheses regarding the ratio of actual/potential transpiration rate in relation to the depletion of soil water

In path (a) the transpiration rate is maintained at maximum potential rate until the available water is exhausted; path (b) shows that transpiration begins to decline at some intermediate stage of soil water depletion; in (c) transpiration declines progressively, though not necessarily linearly over the whole range of soil water depletion.

2.2.5.3 MODELS

Despite advances in the understanding of the effects of soil conditions on transpiration, predictions in the field are still made using semi-empirical relationships such as those of Penman (1949) and Zahner (1967). The gist of these models is to fix the limiting values on the moisture content (i.e. 100% and the wilting point) and to select a convenient transition function.

In the United Kingdom a simple model adopted by the agricultural branch of the Meteorological Office for general farming practice, is to assume that potential evaporation is maintained until half the maximum deficit has been attained — at which point the rate of extraction drops sharply to half the potential rate, until reaching the maximum deficit.

A second simple approach which is often incorporated into computer simulation models, e.g. Fleming (1975) is to assume that the reduction is linear and depicted by pathway (c) of fig. 2.5.

A more complex model (though not necessarily more accurate) which has been used extensively for hydrological predictions, is the root constant concept proposed in Penman (1949). The root constant C defines a specified amount of soil moisture (expressed in mm equivalent depth) which can be extracted from the soil without difficulty by a given vegetation, such that $E = E_T$ where $SMD < C$. A further 25 mm of moisture can then be extracted with increasing difficulty; thereafter, extraction becomes minimal ($E \sim 0.1\ E_T$, when $SMD \gtrsim 3C$). Typical figures for the transpiration associated with a root constant of 75 mm are:

Potential (mm)	100	125	150	175	250
Actual (mm)	98	109	112	115	120

and these can be used to define the general shape of the transition function as shown in fig. 2.6.

To allow for different types of vegetation, Grindley (1969) equated the root constants to the maximum values of SMD as shown in table 2.5. The root constants refer strictly to vegetation type and make no allowance for soil type, except in so far as the soil type is reflected in the vegetation which it carries. Where soils are known to be poor, allowance can be made by reducing the root constant.

The success of these generalised schemes depends on the assumption that the way in which soil moisture depletes within two extremes cannot vary much (Penman, 1967). Also, it is probable that their overall accuracy is commensurate with the accuracy of determining the potential rate. Future developments are likely to be based on the Penman-Monteith equation by exploiting correlations between the surface resistance and the soil moisture status, such as those shown in Russell (1980).

Table 2.5 Correspondence between root constant and maximum soil water deficit tolerated by plants

Maximum soil moisture deficit (mm)	Root constant (mm)
250	200
200	140
150	97
125	75
100	56
50	13

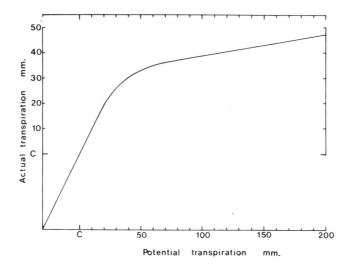

Fig. 2.6 Penman's drying curve (based on data from Grindley, 1969)

2.2.6 Calculation of soil moisture deficit

Knowledge of the soil moisture deficit is an important factor in planned irrigation. To serve such purposes regular estimates of the soil moisture deficit are prepared for distribution by the U.K. Meteorological Office; the estimates, based on climate and vegetation, were originally intended as aids to authorities for flood warning — since the risk of flooding after heavy rain was considered to increase sharply once the 'cushioning' effect of the vegetation has been eliminated.

Table 2.6   Calculation of deficits for catchment area

Month	P_n (mm)	E_T (mm)	$P_n - E_T$ (mm)	Cumulative potential deficit (mm)	Actual** deficit C_{200}	C_{75}	Catchment area (mm)
April	80	30	+50	0*	0	0	0
May	50	79	−29	29	29	29	23
June	5	100	−95	124	124	108	91
July	13	100	−87	211	211	117	122

*Excess drains
**From fig. 2.7

Table 2.7   Calculation of replenishment

Month	P_n (mm)	E_T (mm)	$P_n - E_T$ (mm)	Cumulative potential deficit (mm)	Actual** deficit C_{200}	C_{75}	Catchment area (mm)
(a) *First example*							
June	37	108	−71	119	119	107	89
July	115	88	+27	92	92	80	68
August	38	51	−13	105	105	93	78
(b) *Second example*							
June	37	108	−71	119	119	107	89
July	115	88	+27	92	92	80	68
August	0	51	−51	143	143	112	99

The basis of the balance approach for evaluating *SMD* is provided by the equation:

$$P_n - E = \Delta SMD \qquad\qquad [2.17]$$

where P_n is the net rainfall, E the actual evaporation and ΔSMD the change in the soil moisture deficit. This approach, originally developed in Penman (1949) is illustrated in tables 2.6 and 2.7 in conjunction with fig. 2.6. In these tables, C_{75} refers to short-rooted vegetation with a root constant 75 mm and C_{200} is the corresponding value for long-rooted vegetation. The basis of the approach is to calculate the cumulative potential deficit (assumed to be zero initially) using $P_n - E_T$ and to convert this to the actual deficit for a particular root constant using fig. 2.6. For example, if the potential deficit is 211 mm, then for short-rooted vegetation this is written $211 = 75 + 136$. Reading along the horizontal axis to the position $C_{75} + 136$ the corresponding value on the vertical axis is $C_{75} + 42$, i.e. the actual soil moisture deficit is $75 + 42 = 117$ mm. Where the potential deficit is less than the root constant, water is assumed to be freely available and the actual deficit is identical to the potential deficit.

When rainfall exceeds evaporation, the difference is deducted from the existing soil moisture deficits for each zone until $SMD = 0$. Beyond this, any excess precipitation percolates to the permanent groundwater or runs off.

A special case arises when $E > R$ after a period of decreasing soil moisture deficit. The mechanism of replenishment is that the soil is replenished from the top downwards and it is assumed that such excess water is within easy reach of the root system. Two cases then arise as shown in table 2.7; table 2.7 (a) shows the case when the excess $E - P_n$ $(E > P_n)$ in the later period does not exceed the magnitude $P_n - E$ $(P_n > E)$ in the earlier; here the depletion of the moisture reserves continues at the potential rate. Table 2.7 (b) illustrates the converse situation when $|E - P_n|$ in the later period exceeds $|P_n - E|$ in the earlier period; in this case the depletion of the soil moisture reserves is curtailed after the consumption of excess rainfall in the upper layers of the soil.

In tables 2.6 and 2.7 the final column is based on a weighted average taking account of the different land forms within an area. In the example, it is assumed that 50% of the area is covered by short-rooted vegetation (C_{75}), 30% is covered by long-rooted vegetation (C_{200}) and 20% of the area is riparian, i.e. the permanent groundwater table is so near the surface that moisture is always freely available. With these assumptions and considering the month of July in table 2.6 the average deficit over the catchment is $(117 \times 0.5) + (211 \times 0.3) + (0 \times 0.2) = 122 \times 1.0$ as shown in the final column. Further discussion of this approach and its use for preparing maps of the soil moisture deficit over the United Kingdom is given in Grindley (1967, 1969, 1972).

2.2.7 *Interception*

Vegetation may intercept water falling in the form of snow, hail, rain or overhead irrigation as well as droplets of water in low cloud, mist and fog. In the early stages of interception by a dry canopy, much of the water is retained. There appears, however, to be a fairly well defined storage capacity for any given canopy; when this is exceeded further intercepted water either drips from the canopy or runs down the stems. Water reaching the ground by the combination of through-fall and stemflow is the net precipitation P_n. Interception refers to the difference between the gross precipitation P_g and P_n, and corresponds to the stored water which is evaporated directly from plant surface. Interception loss, I, is often considered to have two components: evaporation during the storm; and water remaining at the end of the storm, and evaporated subsequently.

There have been numerous studies of interception in forests, but relatively few in short vegetation. Most of the results are expressed in the form of the empirical regression equation:

$$I = a P_g + b \qquad\qquad [2.18]$$

where a and b are regression coefficients and P_g the gross precipitation. This type of equation has application for individual storm events and should not be used for periodic (weekly, monthly) summations of gross precipitation. The review of Zinke (1967) for forests in the United States, contains examples of this approach; similarly, Delfs (1955) and Molanchov (1963) provide data for European forests. Corresponding information for grasses and crops is given in Dunne and Leopold (1978).

One of the drawbacks of empirical models of the type shown by the equation is that they give no indication of the mechanisms controlling interception, and take no account of climate factors — especially when it is recognised that interception is a manifestation of evaporation. There are also questions about the precise significance of interception as a mode of water loss. These arise because transpiration ceases when foliage is thoroughly wet. Thus while water may be deemed as 'lost' (in terms of the soil moisture reserves) because of direct evaporation from plant surfaces, there is an associated 'saving' because soil water is no longer withdrawn to support transpiration. To appreciate these aspects it is necessary to examine the process using physically based models.

2.2.7.1 PHYSICALLY BASED REGRESSION MODEL

For a storm large enough to saturate a canopy, the interception loss can be considered as comprising three components: losses during the wetting period (duration t') prior to saturation; evaporative losses from the saturated canopy during the storm; and finally, the loss from storage of capacity S after the storm has ceased, as shown in equation [2.19].

$$I = \underset{\text{wetting}}{\int_{0}^{t'} E\, dt} + \underset{\substack{\text{saturated} \\ \text{canopy}}}{\int_{t'}^{t} E\, dt} + \underset{\text{storage}}{S} \qquad [2.19]$$

Gash (1979) analysed the transient terms of equation [2.19] using a theory proposed in Rutter *et al.* (1972); it was shown that if the meteorological conditions prevailing during any wetting-up period are sufficiently similar to those prevailing for the rest of the storm, then equation [2.19] transforms into equation [2.18] with coefficients identified by:

$$a = \frac{\overline{E}}{\overline{R}}, \text{ and } b = \frac{\overline{R}S}{\overline{E}}\phi\, \ln\phi \quad \text{where } \phi = 1 - \frac{\overline{E}}{\overline{R}}\frac{1}{1 - p - p_t} \qquad [2.20]$$

In these identities: $\overline{E}$ is the mean evaporation rate from a saturated canopy during rainfall; $\overline{R}$, the corresponding mean rainfall rate; p, the free throughfall coefficient, i.e. the proportion of the rain which falls through the canopy without striking a surface; and p_t, the proportion which is diverted to trunks as stemflow (usually very small).

Where single values of a and b are derived from a regression involving several storms, their values are considered representative of any individual storm in the set. There is also the implicit assumption that the ratio $\overline{E}/\overline{R}$ is a constant and highlights the physical basis of the coefficient 'a', but does not imply that $\overline{E}$ and $\overline{R}$ are mutually dependent. Considering the coefficient b, a series expansion of the function $\phi\, \ln\phi$ for the limit $\phi \to 1$ yields the approximation:

$$b \sim \frac{S}{1 - p - p_t} \qquad [2.21]$$

which indicates the dependence of b on the canopy storage capacity.

As a development of this analysis Gash (1979) assumed that the real rainfall pattern could be envisaged as a series of discrete storms; this generates a model for the integrated loss, which depends principally on rainfall data and on an estimate of $\overline{E}$. For tall crops, and neglecting evaporation from the trunks, it takes the form:

$$I = P_s + \frac{\overline{E}}{\overline{R}} (P_g - P_s) + nS \qquad [2.22]$$

in which P_s is the amount of precipitation in rain storms less than $S/(1 - p)$; n, the number of storms with precipitation greater than $S/(1 - p)$; and other variables are as previously defined. Provided there is knowledge of parameters p and S, the description requires knowledge of the rainfall pattern together with the terms $\overline{E}$ and $\overline{R}$ which may be regarded as climatological entities varying from one broad climatic region to the next.

2.2.7.2 FACTORS AFFECTING INTERCEPTION LOSS

Although the developments of Gash (1979) provide a deeper understanding of the regression approach, there are still many factors which are difficult to reconcile and these will be considered below. For a fuller discussion, the reader is referred to Blake (1975) or Rutter (1975).

Rate of evaporation of intercepted water

Figure 2.4 illustrated the dependence of the evaporation rate on the resistances r_s and r_a and indicated that the reduction $r_s \rightarrow 0$ (by surface wetting) increases the evaporation rate. The ratio of the evaporation rate under wet (E_i) and dry (E_t) conditions can be obtained from equation [2.13]; by assuming that the available energy is identical for the two states, this leads to the relationship:

$$\frac{E_t}{E_i} = \frac{\Delta/\gamma + 1}{\Delta/\gamma + (1 + r_s/r_a)} \qquad [2.23]$$

Values of Δ/γ can be obtained from table 2.2 and the remaining critical parameter is the ratio r_s/r_a. For grass and short crops $r_s/r_a \sim 1$, whereas for trees $r_s/r_a \gtrsim 10$. This shows that the evaporation rate from short wet vegetation is roughly twice as great as the transpiration rate in dry conditions, whereas $E_i \sim 10\,E_t$ for trees.

Canopy storage capacity

This refers to the amount of water stored on the saturated canopy when rainfall and throughfall have ceased. Generally the denser the foliage the greater the storage, and this depends on age and season. The form of leaves is also important, e.g. on broad leaves, drops of rain run together, form large droplets and drop off, whereas on coniferous trees the drops are held apart and do not drop off so easily. Data quoted in Rutter (1975) for various vegetation types indicates a storage capacity S ranging from 0.4 mm to 2.0 mm; there was no strong indication of a separation of range between forest and herbaceous communities, although trees in full leaf generally possess a slightly greater storage capacity.

Rainfall intensity and duration

If a storm is defined as a period of rainfall interrupted by rainless periods not exceeding some arbitrary duration, say 3 hours, water for evaporation may be available throughout the period and the storage capacity may be partially depleted by evaporation and then recharged by rain several times. The size of the storage capacity in relation to the duration of breaks in the storm and the evaporation during these periods, will be significant in determining the interception loss − being more important when a given depth of rain falls intermittently than when it is continuous (see Rutter, 1975).

Table 2.8   Details of annual water balance from two forests in Great Britain based on data from Calder (1979) and Shuttleworth and Calder (1979)

	Gross rainfall (mm)	*Inter-ception* (mm)	*Trans-piration* (mm)	*Total evaporation* (mm)	*Rainfall duration* (h)	*Potential evaporation* (mm)
Thetford	595	213	353	566	350	643
Plynlimon*	2260	560	292	852	1380	500

*Estimates for 1975

The importance of rainfall climate is illustrated in table 2.8 which compares the annual precipitation and evaporation losses from Thetford forest and a forest at Plynlimon which are located in regions of Britain with low and high rainfall respectively. The data for Thetford Forest shows that interception accounts for 38% of the total evaporative loss whereas at Plynlimon the magnitude of the interception loss is almost three times greater, and represents 66% of the total evaporative loss. A comparison of the potential evaporation for the two areas shown that potential evaporation is less at Plynlimon than for Thetford; it reinforces the view that the differences in the interception loss are primarily due to the different durations of rainfall, and hence of the canopy wetness in the two areas. The larger discrepancy between the total evaporation and the potential evaporation for Plynlimon, and the similarity at Thetford also point to difficulties in using the Penman method for estimating evaporation from trees.

2.2.7.3 INTERCEPTION AND REGIONAL EVAPORATION

In an attempt to include the effect of interception in estimates of regional evaporation, Thom and Oliver (1977) argued that wetting periods reduce the average surface resistance in dry conditions (r_{sd}) by an amount depending on the fraction of the month during which the surface is wet; from this, the effective surface resistance may be stated as:

$$r_s = (1 - I/E)\, r_{sd} \qquad\qquad [2.24]$$

where I is that part of the total evaporation (E) occurring directly from intercepted precipitation. By substituting the above expression into equation [2.13], Gash (1978) indicated that the long-term evaporation could be represented by:

$$E = E_t + I(1 - C_t) \qquad\qquad [2.25]$$

where E_t is a physically realistic transpiration estimate and I, the interception loss; the term C_t is the ratio of the evaporation under total dry conditions to the evaporation under total wet conditions (see equation [2.23]) and takes account of the fact that the transpiration estimate is also made while the evaporation of intercepted water is in progress.

Calder and Newson (1979) pursued this approach and proposed a 'broad brush' model which focused on the impact of interception losses on the water yield from forested catchments in the British uplands. Their model was based on the following observations: first, that interception losses may be expressed as a fraction of the annual precipitation; second, that *the Penman* E_T *estimate (equation [2.12]) is generally accepted as a reliable method for estimating annual evaporative losses (transpiration plus interception losses) from grass and other short crops*; third, that Penman's E_T also provides an approximate estimate of *transpiration* from spruce forest; and last, that transpiration does not occur whilst forest canopies are wet.

These observations provide the basis of a simple evaporation model — suggesting that the evaporation loss from an upland catchment may be estimated from the sum of the proportional losses arising from the forested and non-forested areas, thus:

annual evaporation = fraction of the catchment under grass X
grass annual evaporation + fraction of
the catchment under canopy coverage X
(forest annual transpiration + forest
annual interception loss)

$$= (1 - f) E_T + f[(1 - w) E_T + \alpha P]$$
$$= E_T + f(\alpha P - w E_T) \qquad [2.26]$$

where f is the fraction of the catchment area with complete canopy coverage, P is the annual precipitation (mm), α is the interception fraction and w the fraction the year when the canopy is wet. At Plynlimon in central Wales, w was found to be 1.5 times the fractional number of rain hours per year and may be expressed as:

$$w = \frac{\text{annual precipitation (mm)}}{\text{mean rainfall intensity (mm h}^{-1})} \times 1.71 \times 10^{-4} \qquad [2.27]$$

Calder and Newson (1979) stressed the following restrictions on the model's use: (a) it should not be applied on a time scale of less than a year; (b) it is not designed for drier lowland areas; (c) it is inappropriate where snow is a major proportion of the annual precipitation; and (d) that non-forested parts of the catchment are covered by short vegetation.

Example 2.3 Interception model I

From a review of the literature on interception in hardwood forests in the eastern United States, Helvey and Patric (1965) concluded that the following formulae relating throughfall (T) to stemflow (St) to gross rainfall (P_g) and season are applicable:

Growing season (leaves present)

$$T = 0.901 \, P_g - 0.787 \qquad \text{mm}$$
$$St = 0.041 \, P_g - 0.127 \qquad \text{mm}$$

Dormant season

$$T = 0.914 \, P_g - 0.381 \qquad \text{mm}$$
$$St = 0.062 \, P_g - 0.127 \qquad \text{mm}$$

Calculate the net rainfall and the interception loss during a sequence of four storms applying 50 mm rainfall each and a sequence of 40 storms supplying 5 mm each:

	T (mm)	St (mm)	$P_n = T + St$ (mm)	$I = P_g - P_n$ (mm)	I/P_g (%)
Growing season					
50 mm storms	177	8	185	15	8
5 mm storms	149	3	152	48	24
Dormant season					
50 mm storms	181	12	193	7	4
5 mm storms	168	7	175	25	13

From these calculations, it is observed that the largest interception losses occur during the growing season and are associated with a sequence of small storms, rather than infrequent large storms.

Example 2.4 Interception model II

A pine forest with a storage capacity 0.8 mm of rainfall is located in a region in which the mean evaporation rate from the saturated canopy is 0.19 mm h^{-1}. Analysis of data over a period during which the total rainfall was 896 mm indicated 132 storms with rain sufficient to saturate the canopy ($\sim$ 1.3 mm from dry conditions). During these storms, the accumulated rainfall was 822 mm and the associated mean rainfall rate was 1.38 mm h^{-1}. Estimate the interception loss.

Rain during storms $\gtrsim 1.3$ mm $(P_g) = 822$ mm
Rain during storms < 1.3 mm $(P_s) = 896 - 822$
$$= 74 \text{ mm}$$

From equation [2.22]

$$I = 74 + \frac{0.19}{1.38}(896 - 74) + 132 \times 0.8$$
$$= 293 \text{ mm}$$

This indicates that roughly 33% of the incident rainfall is lost by interception.

2.3 Infiltration

Infiltration is defined as the movement of water through the soil surface into the soil profile under the influence of gravity and capillarity. The infiltration rate is the actual rate at which water enters the soil profile; it has a limiting value termed the infiltration capacity, representing the maximum rate at which rainfall can be absorbed by a soil in a given condition.

Infiltration is highly desirable agriculturally; and hydrologically is important in determining the disposition of precipitation falling upon a catchment area. If the rainfall intensity is lower than the infiltration capacity of the soil, all the falling rain not held as surface storage will infiltrate into the soil. Otherwise, runoff may occur — the rate being determined by an imbalance between the rainfall rate and the infiltration capacity.

2.3.1 Infiltration process

Infiltration involves three interdependent processes: entry through the soil surface; depletion of available storage capacity; and transmission through the soil. Inspection of fig. 2.7 shows that the infiltration rate declines rapidly during the early part of a storm and reaches an approximately constant value after one or two hours of rain.

Many factors influence the shape of the infiltration capacity curve, but the most important characteristics are: rainfall characteristics, soil properties, vegetation and land use. Coarse-textured soils, e.g. sands, possess greater permeability than soils with fine texture, e.g. clay. Vegetation and land use may affect the infiltration of water in various ways: roots perforate the soil keeping it unconsolidated and porous; decaying organic matter promotes a crumb structure and improves permeability; surface litter (unincorporated organic matter) absorbs the kinetic energy of incident raindrops and reduces surface compaction. The presence of surface litter — particularly in forests — is a valuable means of maintaining 'good' soil conditions, i.e. aiding infiltration; its removal, perhaps by burning, is generally detrimental to the soil structure and can lead to significant reductions in the infiltration rate.

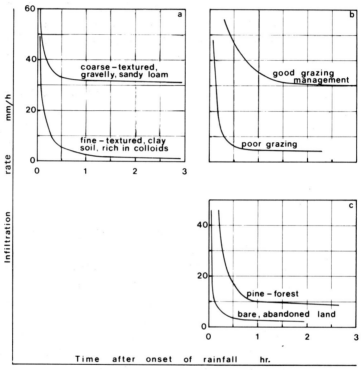

Fig. 2.7 Infiltration rate curves for soils (a) of different textures; (b) with different vegetative covers; and (c) with different land use practices

A prime agency of soil-compaction damage in land use is the use of machinery. Track-type tractors, such as those used in logging, exert a pressure of ~ 48 kN m^{-2} (7 p.s.i.) depending upon vehicle weight and track width. Loading from wheeled vehicles, e.g. trucks and cars, is even greater. Trampling by animals and people can also create significant compaction damage. Compaction results in a reduction of pore space (particularly macropore space); an increased bulk density; and in reduced permeability. Soils of equivalent texture compact to a greater extent when in a wet condition than when they are dry.

2.3.2 Infiltration models

Infiltration theory is, in principle, well understood and has been described in detail by Philip (1969), Hillel (1971) and many others. However the techniques have not as yet been adapted for general application in hydrology — perhaps reaching their greatest use in catchment modelling (discussed in Fleming and Smiles, 1975). In practice, it is more feasible to estimate the volume of runoff by applying a coefficient to the volume of precipitation and to distribute this volume in time by applying the principles of the unit hydrograph (see section 2.4.4 and example 2.6). Retrospectively, the analysis of the runoff hydrograph from plots and watersheds has provided valuable insight into the infiltration characteristics of soils and arrays of land use.

Of particular interest is the empirical infiltration model developed by Holtan et $al.$ (1967); this was based on field experience, on the premise that soil moisture storage, surface-connected porosity and the effect of root paths are the dominant factors influencing the infiltration rate. The normal form is:

$$v_t = a(S_p - i_t)^n + v_c \qquad\qquad [2.28]$$

where a and n are empirical constants; S_p is the potential storage in the root zone at storm commencement, i.e. $(\theta_s - \theta_r)d$, with θ_s as the soil water content at saturation, θ_r the uniform initial soil water content at the beginning of the rainfall event, and d the root depth (Body, 1975); i_t is the volume infiltration to time t; and v_c the final infiltration rate after prolonged wetting. The exponent n usually has a value of 1.4, and a is rationalised — related to the manner in which crop roots enhance the soil permeability as illustrated in table 2.9. In some versions, e.g. Holtan and Lopez (1971), a is modified by a seasonal growth index. The U.S. Department of Agriculture has established v_c for most major soil types in the U.S.A. as described in Musgrave and Holtan (1964) (see table 2.14). Though equation [2.28] cannot be discussed in terms of infiltration theory, it is in extensive use in the United States and is incorporated into the U.S.D.A. Hydrograph Laboratory Hydrologic Model (Holtan and Lopez, 1971; Holtan et $al.$, 1975).

Table 2.9 is of particular interest, because it ranks the various land use practices in order of their influence on the infiltration capacity of various soils, and serves as a useful guide to the planner in identifying the probable infiltration character of areas within a drainage basin.

Table 2.9   Tentative estimates of vegetation parameter a in equation [2.28]

Land use or cover	Basal area rating* ($mm\ h^{-1}$ per $mm^{1.4}$)	
	Poor condition	Good condition
Fallow**	0.027	0.082
Row crops	0.027	0.055
Small grains	0.055	0.082
Hay (legumes)	0.055	0.110
Hay (sod)	0.110	0.165
Pasture (bunchgrass)	0.055	0.110
Temporary pasture (sod)	0.110	0.165
Permanent pasture (sod)	0.219	0.274
Wood and forests	0.219	0.274

*Evaluated at plant maturity as the percentage of the ground surface area occupied by plant stems or root crowns; adjustments are required for 'weeds' and 'grazing'.

**For fallow land only. 'Poor condition' means 'After row crop' and 'Good condition' means 'After sod'.

Table 2.10   Soil infiltration rates

Soil type	Good structure ($mm\ h^{-1}$)	Poor structure* ($mm\ h^{-1}$)
Coarse sand	19–25	13
Fine sand	13–19	9
Fine sandy loams	13	8
Silt loams	10	7
Clay loams	8	6

*'Structure' refers to the degree of compaction and clogging in the surface layers

A simplified approach described in Wiesner (1970) and useful in irrigation practice, is based on the data shown in tables 2.10 and 2.11; these show that the infiltration rate is estimated by the product of the soil intake rate and its cover factor. Thus, on a fine sandy soil of good structure and with a good grass cover, infiltration rates are probably between 13×2.0 and 13×3.0 mm h^{-1}.

Table 2.11   Cover factors for adjusting values of infiltration rates for bare soils

Type (1)	Condition of cover (2)	Factor (3)
Permanent (forest and grass)	Good	3.0 – 7.5
	Medium	2.0 – 3.0
	Poor	1.2 – 1.4
Close growing crops	Good	2.5 – 3.0
	Medium	1.6 – 2.0
	Poor	1.1 – 1.3
Row crops	Good	1.3 – 1.5
	Medium	1.1 – 1.3
	Poor	1.0 – 1.1

Notes

Column (2): Terminology defined in American Society of Civil Engineers
(1949), but broadly consistent with classification in table 2.13

Column (3): Ratio of infiltration rate for a given soil with cover to the
corresponding rate for the same soil without cover

2.4 Runoff

2.4.1 Perspective

When soil becomes saturated or the precipitation rate exceeds the
infiltration rate, surface detention of water begins to occur filling
small surface irregularities, plough furrows, etc. and may eventually
spill to yield an overland flow, i.e. a flow across the land surface
towards stream channels. Subsurface flows may also occur with
lateral flows through the upper soil horizons towards streams (termed
subsurface runoff) and with vertical flows towards the water table;
the latter is referred to as deep percolation, the water moving slowly
into streams and lakes and supporting runoff during dry weather. In
total these flows form a portion of the precipitation which is referred
to as 'runoff', though if the term is used alone it usually refers to
surface runoff.

From the hydrologic point of view, the runoff from a basin can be
considered as being influenced by two major groups of factors,
climatic and physiographic.

Climatic factors include the form of precipitation, interception,
evaporation and transpiration, all of which exhibit seasonal variations.

Physiographic factors can be subdivided into basin characteristics and channel characteristics. Basin characteristics refer to geometric factors such as size, shape and slope combined with physical factors including land use, cover, soil type, geological and topographic features; channel characteristics are mostly related to the hydraulic properties of the channel which govern stream flow (see section 2.6).

In the management of runoff we need to know flow volumes for estimating reservoir size; peak rates for the design of spillways, flood banks and conveyance channels; and low rates for estimating the yield of diversions and navigation channels or for supporting abstraction, fishing and other amenity uses.

In many situations the controls of runoff are very sensitive to disturbance. For example, in rural areas the removal of forest — perhaps during construction — can generate large amounts of storm runoff, whereas the previous runoff was a slow subsurface percolation. Changes promoted by urbanisation may be even more forceful since the volume of runoff is governed primarily by infiltration characteristics (related to land slope, soil type as well as the type of vegetation cover) and is directly related to the percentage of the area covered by impervious surfaces. With urbanisation the increased storm runoff leads to difficulties in storm drainage controls, stream channel maintenance, groundwater recharge and stream water quality.

Two main approaches exist for the prediction of runoff. The first is based on a statistical analysis of stream flow measured at gauging stations, and the second consists of 'rainfall-runoff' models which are linked to the physical conditions of the drainage area. An early example of the 'rainfall-runoff' model is the 'Rational' method (section 2.4.3) which attempts to quantify peak flows in terms of the rainfall intensity and catchment factors. A more advanced development, also used extensively in engineering practice, is the 'unit hydrograph' (section 2.4.4) which provides a workable basis for estimating the shape of the hydrograph from a specified volume of flow in a given time period (see fig. 2.10). By analysing rainfall/runoff events, consistent rules emerge which make it possible to compare runoff volumes with rainfall amounts and conditions of the catchment. The advent of computers has been accompanied by simulation models such as the Stanford Watershed Model (Crawford and Linsley, 1966); these represent the most sophisticated means of generating accurate forecasts of runoff and the effects of alterations in land use, and are discussed in Fleming (1975).

Runoff spans the entire regime of flow — its prediction requiring fairly comprehensive analysis, such as described in Chow (1964) and Natural Environment Research Council (1975). However, planners often need to make rough and ready calculations of the probable flood magnitudes for small catchments, or for assessing the potential hydrological changes arising from alterations in land use. The methods described below represent a first step and may give some indication of whether a more sophisticated analysis is required. Emphasis is placed on the effects of vegetation and land use. Readers are referred to Chow (1964), Gray (1970) or Dunne and Leopold (1978), or other hydrological texts, to gain a better insight into the way these approaches are used in practice.

2.4.2 Estimation of short-term catchment yield

When evaluating catchment water yields it is important to distinguish between the runoff volume from individual storms and the net run-off over a prolonged period. In the former case a first order estimate can be obtained by subtracting the antecedent soil moisture deficit from the storm rainfall:

$$R = P_g - SMD \qquad (P_g \geqslant SMD) \qquad\qquad [2.29]$$

where SMD is calculated from a daily balance between the daily rainfall and the actual evaporation from the catchment following the principles discussed in section 2.2.6. For more accurate forecasts, account must be taken of the balance between the rainfall intensity and the infiltration rate as well as the various storage values. Over prolonged periods, and in catchments underlain by an impervious layer, the catchment yield is essentially a balance between the precipitation input and the evaporative losses (equation [2.2]) – the latter being estimated using the methods described in section 2.2.

One of the most widely tested approaches for determining the short-term runoff arising from small catchments from a given sequence of daily rainfall is that of the U.S. Soil Conservation Service (1972). In this procedure the catchment response to rainfall is estimated from the empirical relationship:

$$Q = \frac{(P - I_a)^2}{(P + 4I_a)} \qquad\qquad [2.30]$$

where Q is the volume runoff, P is the rainfall and I_a is the initial abstraction or initial loss. The initial abstraction consists of interception losses, surface storage and water which infiltrates the soil prior to runoff, and is estimated initially by:

$$I_a = 0.2\,S \qquad\qquad [2.31]$$

where S is the maximum potential difference between rainfall and runoff in inches, starting at the storm's beginning. For convenience in evaluating antecedent moisture, soil conditions, land use and conservation practices, the U.S. Soil Conservation Service (1972) defines:

$$S = \frac{1000}{N} - 10 \qquad\qquad [2.32]$$

where N is an arbitrary curve number varying from 0 to 100. Thus if $N = 100$, $S = 0$ and $Q = P$.

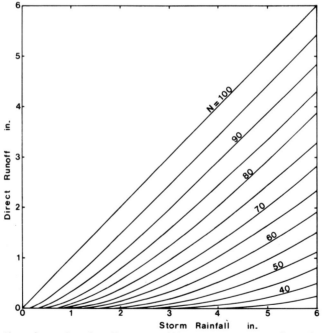

Fig. 2.8 Chart for estimating direct runoff based on equations [2.30] to [2.32] and referring to tables 2.12 to 2.16

Curve numbers (see fig. 2.8) can be obtained for various soil-cover complexes for average soil moisture conditions using tables 2.12 to 2.14. When it is necessary to make runoff estimates for below average

Table 2.12 Runoff curve numbers for hydrologic soil-cover complexes under average conditions of antecedent moisture (class II in table 2.15)

Land use or cover	Treatment or practice	Hydrologic* condition	Hydrologic** soil group A	B	C	D
Fallow	Straight row	Poor	77	86	91	94
Row crops	Straight row	Poor	72	81	88	91
	Straight row	Good	67	78	85	89
	Contoured	Poor	70	79	84	88
	Contoured	Good	65	75	82	86
	Contoured and terraced	Poor	66	74	80	82
	Contoured and terraced	Good	62	71	78	81
Small grain	Straight row	Poor	65	76	84	88
	Straight row	Good	63	75	83	87
	Contoured	Poor	63	74	82	85
	Contoured	Good	61	73	81	84
	Contoured and terraced	Poor	61	72	79	82
	Contoured and terraced	Good	59	70	78	81
Close-seeded legumes or rotation meadow	Straight row	Poor	66	77	85	89
	Straight row	Good	58	72	81	85
	Contoured	Poor	64	75	83	85
	Contoured	Good	55	69	78	83
	Contoured and terraced	Poor	63	73	80	83
	Contoured and terraced	Good	51	67	76	80
Pasture or range		Poor	68	79	86	89
		Fair	49	69	79	84
		Good	39	61	74	80
	Contoured	Poor	47	67	81	88
	Contoured	Fair	25	59	75	83
	Contoured	Good	6	35	70	79
Meadow (permanent)		Good	30	58	71	78
Woodlands (farm woodlands)		Poor	45	66	77	83
		Fair	36	60	73	79
		Good	25	55	70	77
Farmsteads			59	74	82	86
Roads, dirt			72	82	87	89
Roads, hard-surface			74	84	90	92

*Refer to table 2.13
**Refer to table 2.14

(dry) and above average (wet) conditions, representative N values can be approximated using the additional tables 2.15 and 2.16. The antecedent moisture levels are classified on the basis of the total precipitation occurring within the preceding five days. If a storm continues for several days, the rainfall should be broken into daily totals and the antecedent moisture class modified accordingly.

Table 2.13   Classification of vegetative covers by their hydrologic properties

Vegetative cover	Hydrologic condition
Crop rotation	Poor: Contain a high proportion of row crops, small grains, and fallow
	Good: Contain a high proportion of alfalfa and grasses
Native pasture or range	Poor: Heavily grazed or having plant cover on less than 50% of the area
	Fair: Moderately grazed; 50-75% plant cover
	Good: Lightly grazed; more than 75% plant cover
	Permanent meadow: 100% grass cover
Woodlands	Poor: Heavily grazed or regularly burned so that litter, small trees, and brush are destroyed
	Fair: Grazed but not burned, there may be some litter
	Good: Protected from grazing so that litter and shrubs cover the soil

Table 2.14   Classification of soils after a period of prolonged wetting when planted with row crops

Group	Runoff and infiltration characteristics	Soil characteristics
A	Low runoff potential, high infiltration rates (8-12 mm h^{-1})	Deep sands, deep loesses, aggregated soils
B	Moderate infiltration rates (4-8 mm h^{-1})	Shallow loess and sandy loams
C	Slow infiltration rates (1-4 mm h^{-1})	Many clay loams, shallow sandy loams, soils low in organic matter and soils high in clay
D	High runoff potential, very slow infiltration rates (0-1 mm h^{-1})	Soils of high swelling percentage, heavy plastic clays and certain saline soils

Table 2.15   Rainfall limits for estimating antecedent moisture conditions

Antecedent moisture condition class	5-day total antecedent rainfall (in)	
	Dormant season	Growing season
I	Less than 0.5	Less than 1.4
II	0.5 – 1.1	1.4 – 2.1
III	Over 1.1	Over 2.1

Table 2.16   Conversion of runoff curve numbers (N) for antecedent moisture condition II to those for conditions I (dry) and III (wet)

N for antecedent moisture condition	N for antecedent moisture condition	
II	I	III
100	100	100
95	87	98
90	78	96
85	70	94
80	63	91
75	56	88
70	51	85
65	45	82
60	40	78
55	35	74
50	31	70
45	26	65
40	22	60
35	18	55
30	15	50
25	12	43
20	9	37
15	6	30
10	4	22
5	2	13

Where a catchment consists of widely differing soil-cover complexes the total runoff can be estimated by calculating the runoff from individual land use types and weighting them according to their proportion of the catchment area (see example 2.6). Where soil-cover complexes are more similar in character, a weighted curve number can be used on the same basis as equation [2.35].

2.4.3 The Rational method

The Rational runoff method is often used for basins up to 25 km² in rural areas and is a widely accepted approach for designing storm sewers. The method assumes that a rainstorm of uniform intensity covers the whole basin. Runoff will increase as water from more and more distant parts of the catchment reaches the outlet. When the whole drainage area is contributing, a steady state is reached and the discharge reaches a maximum. The time required to reach this steady state is called the time of concentration of the basin and after this the peak runoff rate Q_{pk} is:

$$Q_{pk} = CIA \hspace{5cm} [2.33]$$

where C is the runoff coefficient, I the rainfall intensity (in h⁻¹) and A is the drainage area (acres). In these units the peak runoff is calculated in c.f.s. However if metric units are to be used equation [2.33] becomes

$$Q_{pk} = 0.278\ CIA \hspace{4cm} [2.34]$$

where Q_{pk} is in m³ s⁻¹, I is in mm h⁻¹ and A is in km².

Table 2.17   Values of the Rational runoff coefficient C

Soil type	Watershed cover		
	Cultivated	Pasture	Woodlands
Sandy and gravelly soils	0.20	0.15	0.10
Loams and similar soils without impeding horizons	0.40	0.35	0.30
Heavy clay soils or those with a shallow impeding horizon; shallow soils covering bedrock	0.50	0.45	0.40

The runoff coefficient depends on catchment characteristics and expresses some of the storage effects of a catchment in attenuating flood peaks. The coefficient C also includes the effects of infiltration and tends to be higher for storms with longer return periods (i.e. higher intensities). Many texts, e.g. Dunne and Leopold (1978) list accepted ranges for C which are pertinent to urban and suburban areas. For agricultural areas Schwab et al. (1966) have developed a system for estimating C for various soil-cover combinations and as a rough guide table 2.17 provides insight into their potential influence on the peak runoff rate. It is apparent that peak runoff will be least from woodland areas, this being largely attributable to their superior infiltration character and greater storage capacity within the root zone (see table 2.9).

Where a catchment consists of a variety of land use practices and soils, C can be weighted according to the proportion of the area under cultivation (w_i) and the associated runoff coefficient C_i:

$$\bar{C} = \Sigma w_i C_i \qquad\qquad [2.35]$$

In urban areas, the flood discharge depends on: the percentage of impervious area in a catchment; the percentage of the area served by storm sewerage; and the recurrence interval of the storm (see Leopold 1968). A system for estimating C is discussed in Rantz (1971).

The appropriate rainfall intensity (I) is chosen with reference to the recurrence interval of the storm to be designed for. For example in the British Isles short-duration storms may be roughly estimated using the Bilham formula (see British Standard CP 2005: 1968).

$$I = \frac{14.2}{t^{0.72}}\ F^{0.28} - \frac{2.54}{t} \qquad (\text{mm h}^{-1}) \qquad\qquad [2.36]$$

where F = frequency of occurrence (y)
    t = storm duration (h)
However, more accurate estimates are obtained directly from rainfall records for a particular area — a service provided by the Meteorological Office within the United Kingdom.

The duration of the design storm is taken as the time of concentration of the basin, i.e. *the time required for overland and channel flow to reach the basin outlet from the hydraulically most distant part of the catchment.*

The time of concentration can be estimated in a variety of ways, e.g. a formula referred to in U.S. Soil Conservation Service (1972) is:

$$t_c = \frac{L^{1.15}}{7700 H^{0.38}} \qquad\qquad [2.37]$$

in which t_c (h) depends on L (ft), the length of the catchment along the main stream from the basin outlet to the most distant ridge and H (ft) is the difference in elevation between the basin outlet and the most distant ridge.

As an independent check it is wise to estimate t_c from estimates of the velocities of overland flow and channel flow. The timescale of overland flow depends on the overland travel distance and the slope, together with surface features influencing infiltration and the hydraulic surface roughness; estimates of the overland flow time can be obtained from fig. 2.9; channel velocities may be computed from the Manning formula (see section 2.6.1). Generally, the evaluation of t_c is fraught with difficulty; it is important, because it should match and determine the duration of the design storm. If the duration of a rainstorm does not equal or exceeds the concentration time, the Rational method will overestimate the flood peak (Dunne and Leopold 1978).

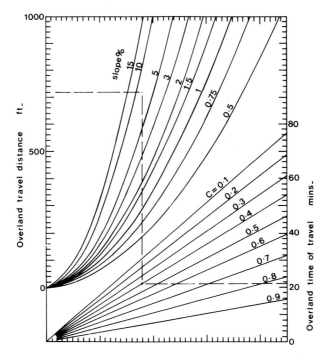

Fig. 2.9 Relation of overland flow time of travel to overland travel distance, average overland slope, and the Rational runoff coefficient (*C*)

Though the Rational method is recognised to have many weaknesses, it is considered to be sufficiently accurate for runoff estimation in the design of relatively inexpensive structures in small catchments where the consequences of failure are limited.

2.4.4 Hydrographs

This refers to a graph showing the discharge, the velocity, or some other property of the water flow as a function of time. It can be regarded as an integral expression of the physiographic and climatic features that govern the relation between rainfall and runoff of a particular drainage basin. A typical hydrograph produced by a concentrated storm event is a single peaked distribution curve of the type shown in fig. 2.10. However multiple peaks may appear on a hydrograph indicating abrupt variation in rainfall intensity, a succession of storms or other causes. In hydrograph analysis, multiple peaked graphs may be separated into a number of single peaked hydrographs.

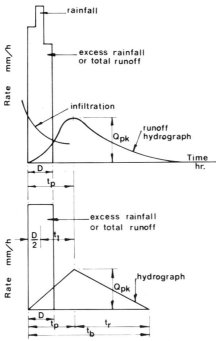

Fig. 2.10 Development of a triangular unit hydrograph. Symbols: Q_{pk} = peak runoff in mm h^{-1}; t_p = time from start of rise to peak rate; t_r = time from peak rate to end of triangle; t_b = time base of hydrograph; D = rainfall excess period; t_l = time from centre of excess rainfall to time of peak

A single peaked hydrograph consists of three parts: a rising limb, a crest segment, and a recession limb. Climatic factors predominate in producing the rising limb whereas the recession limb is largely independent of the storm characteristics producing the runoff, and is heavily dependent on ground conditions. The peak of the hydrograph represents the highest concentration of runoff from a drainage basin. It usually occurs at a certain time after the rain has ended, the time depending on the areal distribution of the rainfall.

Because a storm hydrograph consists of direct runoff (surface runoff and interflow) and baseflow, it is convenient to separate these components as an aid to subsequent analysis. To predict future flood hydrographs, the stormflow is first computed, and then added to the baseflow — which is calculated separately. Here, we shall restrict discussion to storm runoff.

2.4.4.1 UNIT HYDROGRAPH

The unit hydrograph of a basin is the hydrograph of one unit of direct runoff generated by a rainstorm of fairly uniform intensity occurring within a specified period of time. It presents an extremely useful tool for predicting the distribution of storm runoff — especially of peak flows — and for identifying the effects of climatic and physiographic parameters on the runoff process. It is commonly used for catchment areas of up to $\sim$ 5000 km² (2000 square miles). Several studies, e.g. Heerdegen and Reich (1974) have shown that variations in runoff are generally dependent on a few catchment parameters, especially area and the type of vegetal cover; storm parameters seem to have a small effect on the unit hydrograph.

The theory is based in principle on the following criteria:

(a) For a given catchment, runoff producing storms of equal duration will produce surface runoff hydrographs with equivalent time bases regardless of the intensity of the rain.

(b) For a given catchment, the instantaneous discharge from an area will be proportional to the volume of surface runoff for storms of equal duration.

(c) For a given catchment, the time distribution of runoff from a given storm period is independent of precipitation from antecedent or storm periods and reflects all the combined physical characteristics of the basin.

The data required for deriving a unit graph are simultaneous measurements of rainfall and runoff from the basin for a number of years, and also (preferably) some estimate of the infiltration rate. Once a unit graph has been determined for a basin it can be used to obtain the surface runoff hydrographs for storm events on the basin.

2.4.4.2 EFFECTS OF LAND USE

Essential features of the hydrograph may be discussed in terms of the synthetic triangular hydrograph (fig. 2.10) which is useful for predictions in ungauged catchments. From the triangular hydrograph it is evident that:

$$Q_{pk} = \frac{2R}{t_b} \qquad [2.38]$$

where Q_{pk} (mm h^{-1}) is the peak runoff, R (mm) the equivalent depth of runoff and t_b (h) the base length. Inspection of equation [2.38] suggests that any process, such as infiltration, which leads to a reduction in R causes a proportional reduction in the peak flow. Similarly the peak flow is reduced by an extension in the time scale t_b, or the lag time t_p with which t_b has been found to be correlated (see Dunne and Leopold, 1978, p. 343).

From an analysis of flood events occurring in a series of catchments situated in different regions of the Commonwealth of Pennsylvania, Heerdegen and Reich (1974) found that:

$$t_b = 0.60 \, A^{0.2} \, C_w^{0.37} \quad (h) \qquad [2.39]$$

in which A (square mile) is the catchment area and C_w the percentage wooded area. Bell and Songthara om Kar (1969) performed a similar analysis in another series of catchments and found that the lag time could be estimated causing:

$$t_p = AM^{0.33} \quad (h) \qquad [2.40]$$

with A in square miles and M a coefficient which showed a strong relationship with the vegetation cover group as presented in table 2.18. Each of these relationships illustrates the influence of vegetal cover on the runoff process and emphasises the value of woodland in reducing peak flows.

Table 2.18 Values of M in equation [2.40]

Vegetation cover group	Mean M	Standard deviation
Forest and good woodland	2.05	0.50
Good pasture and poor to fair woodland	1.50	0.20
Crops and poor to fair pasture	1.15	0.30
Very poor pasture and desert vegetation	0.60	0.15

Example 2.5 Runoff prediction: Rational method

A watershed has a drainage area of 0.8 km², a length of main channel of 854 m (2800 ft), a difference in elevation (watershed outlet to the most distant ridge) of 46 m (150 ft) and a mixed cover consisting of approximately 15% woodland, 40% pasture and 45% cultivated. The soils are loams without impeding horizons. Determine the peak runoff rate of a 5-year frequency using the observation that the rainfall of duration 11 min with a 5-year frequency is 23 mm.

From equation [2.35] and table 2.17:
$$\overline{C} = 0.15 \times 0.30 + 0.40 \times 0.35 + 0.45 \times 0.40$$
$$= 0.37$$

From equation [2.37]:
$$t_c = 2800^{1.15}/(7700 \times 150^{0.38})$$
$$= 0.18 \text{ h}$$

For the design rainfall:
$$I = 23/0.18$$
$$= 128 \text{ mm h}^{-1}$$

From equation [2.34]:
$$Q_{pk} = 0.278 \times 0.37 \times 128 \times 0.8$$
$$= 10.5 \text{ m}^3 \text{ s}^{-1}$$

Example 2.6 Runoff prediction: Synthetic unit hydrograph

An analysis is to be carried out on a catchment (located in the United Kingdom) shown in fig. 2.11 to determine the runoff arising from a critical storm with a 20-year return period n.b. the information is required for modifying the dimensions of the outfall channel.

Points A, B, C and D were chosen as representative points within sub-areas of the catchment for evaluating the critical timescale for overland flow. Catchment data indicates a runoff volume roughly equivalent to 20% of the incident rainfall and conforms to a synthetic hydrograph of triangular form shown in fig. 2.11.

As a preliminary step, the catchment was inspected and yielded the information summarised in fig. 2.11 and columns (1)–(4) of table 2.19.

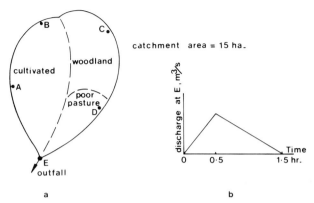

Fig. 2.11 (a) Sketch of land use in a small catchment and (b) the associated synthetic triangular hydrograph at the outfall

Table 2.19   Determination of critical timescale for overland flow

Position (1)	Distance to outfall (m) (2)	Mean slope to outfall (%) (3)	Soil conditions (4)	Rational coefficient (5)	Overland time (min) (6)
A	220	8.0	Loam	0.4	21
B	300	5.0	Loam	0.4	**28**
C	255	7.0	Sandy/loam	0.3	**28**
D	142	15.0	Shallow soil	0.5	12

Notes
Column (5): estimates based on land use and soil type (table 2.17)
Column (6): using fig. 2.9

From table 2.19 the critical time is the maximum time of overland flow i.e. 28 min and is taken as the time of concentration of the flow.

Since the catchment is located in the United Kingdom it is assumed that Bilham's formula is applicable. Substituting $F = 20$ y and $t = 0.47$ h (i.e. 28 min) into equation [2.36] yields the critical rainfall intensity $I = 51.2$ mm h^{-1}.

The incident rainfall has the equivalent depth $It = 51.2 \times 0.47 = 24$ mm.

The runoff volume equivalent

$$R = \underset{\text{precipitation}}{(24 \times 10^{-3})} \times \underset{\text{area}}{(15 \times 10^4)} \times \underset{\substack{\text{catchment} \\ \text{factor}}}{0.2} = 720\text{m}^3$$

From equation [2.38] :

$$Q_{pk} = \frac{2 \times 720}{1.5 \times 3600}$$
$$= 0.27 \text{ m}^3 \text{ s}^{-1}$$

If there is a major stream within the catchment, the procedure for calculating the critical time must be modified to include a component of channel flow time; the latter may be estimated by dividing a channel length (from the assumed point of entry of overland flow) by V, the flow velocity; V can be estimated using the Manning formula (see section 2.6).

Example 2.7 Prediction of short-term catchment yield

A watershed comprises 93 ha of woodland (good condition), 162 ha of pasture (fair condition) and 405 ha of corn (straight row and good rotation). Soils are of type C. After a week without rain during the growing season, a 3-day storm yields 2 in of rain on each day. Estimate the direct runoff for this storm.

Table 2.20 together with the accompanying notes outlines the calculation procedure. The example highlights the response of Q to the cover type and the antecedent soil moisture.

Table 2.20   Calculation scheme for short-term catchment yield

Day (1)	Ante-cedent rainfall (in) (2)	Ante-cedent moisture class (3)	Curve number (N) Corn (4)	Pasture (5)	Wood-land (6)	Runoff Q (in) Corn (7)	Pasture (8)	Wood-land (9)	$\bar{Q}$ (in) (10)
1	0.0	I	70	62	51	0.24	0.09	0.00	0.17
2	2.0	II	85	79	70	0.80	0.52	0.24	0.65
3	4.0	III	94	91	85	1.39	1.16	0.80	1.25

Notes
Column (2): based on previous 5 days e.g. for Day 3, antecedent rainfall =
       0 + 0 + 0 + 2 + 2 in
Column (3): based on column 1 and table 2.15
Columns (4)-(6): N is selected for each soil-cover complex from table 2.12 for
       class II and the equivalent curve numbers for classes I and III
       are obtained from table 2.16
Columns (7)-(9): based on equation [2.30] or fig. 2.8
Column (10): weighted mean e.g. on Day 3

$$\bar{Q} = \frac{1.39 \times 405 + 1.16 \times 162 + 0.80 \times 93}{405 + 162 + 93} = 1.25 \text{ in}$$

2.5 Snow hydrology

Large areas of the earth are snow-covered for at least a portion of the year, and in many countries snow constitutes a major water resource. Snow supplies at least one third of the water used for irrigation in the world (Steppuhn, 1981); in temperate climates major floods occur most frequently in late winter and spring and appear closely associated with snowmelt; its release supports hydro-electric production and urban water supply.

The main influence of vegetation concerns the local distribution of snow and the rate of melting of snow. Much of the associated research has focused on the role of forests in snow hydrology, since grass and low-growing vegetation have a comparatively minor influence on accumulation and melting. However, this does not diminish the importance of low-growing vegetation and tillage practices in securing water supplies in dry regions (see section 5.5 and Steppuhn, 1981).

The ensuing discussion confines itself to snow hydrology in the forest environment. Comprehensive studies on these aspects are *Snow Hydrology* prepared by the U.S. Army Corps of Engineers (1956) and the major Soviet publication on snowmelt by Kuz'min (1961), which deals primarily with conditions on the Russian steppes. Gray and Male (1981) provide a broader insight into snow and its environment—covering principles, processes, management and uses.

2.5.1 Snow accumulation

As might be expected, different surfaces of snow exhibit different abilities to catch and retain snowfall. Kuz'min (1960) has made extensive studies in this area finding, for example, that there is 30-40% greater depth of snow in forested areas than on virgin soil; Rakhmanov (1962) reported similar findings and showed that a greater accumulation takes place within a broad-leafed stand than in coniferous stands. In contrast, many observations, e.g. by the U.S. Army Corps of Engineers (1956) have shown that small forest clearings tend to collect a greater snowpack than under surrounding trees. However as clearings become larger they become more prone to wind-scour; according to Kuz'min (1960) the amount of snow accumulated in a forest clearing depends on the size of the wooded area, e.g. the average depths in clearings were 0.53 m and 0.36 m in areas of 100 m X 200 m and 1 km X 2 km respectively.

Many investigators have found snow accumulation to be inversely related to canopy density. For example the U.S. Army Corps of Engineers (1956) reports a relationship of the form:

$$Y = a - bX \qquad\qquad [2.41]$$

where Y is the snowpack water equivalent expressed in percentage of the water equivalent in the open and X is the canopy density in percentage of complete cover with a and b as constants. Their survey suggested a in the range 95.5-99.9 and b in the range 0.36-0.40; similar findings were reported in Kuz'min (1960).

2.5.2 Prediction of snowmelt

2.5.2.1 TEMPERATURE INDEX METHODS

The prediction of daily snowmelt is a very desirable aid in water management and is also useful in the design of water control structures. Much of the prediction in the past has been based on temperature-index methods which are of the form:

$$M = k (T - T_0) \qquad\qquad [2.42]$$

where T is the measured air temperature; T_0 is a base temperature at which snowmelt is assumed to start (usually taken as 0°C); k is a temperature index and M is taken as the water yield (mm day^{-1}). The assumptions behind this approach are that the temperature index $(T - T_0)$ gauges the energy input and is related directly to stream runoff, and k can be treated as a constant. Coefficient k is a lumped parameter which contains information about the yield of melt water, as well as characteristics of the drainage. Since heat supplied by the air is only one of the sources of heat available for snowmelt at a given point, considerably closer correlations exist when several sources of heat are taken into account. However, despite the approximate nature of this technique, surprisingly good results have been reported in the literature (see Gray, 1970). Temperature index methods are considered to have the best applicability to large forested basins with homogeneous hydrological characteristics; they offer the twin advantages of simplicity and speed in cases where the attainable degree of accuracy is adequate for forecast purposes.

2.5.2.2 THE ENERGY BUDGET METHOD

The rate of melting is governed by the availability of energy to raise the temperature of the snowpack to 0°C and to supply the latent heat of fusion. When melting occurs, the energy balance may be written:

$$M = \frac{Q_\mathrm{m}}{\lambda_\mathrm{f}\rho} \qquad\qquad [2.43]$$

where M is the rate of release of melt water (m s^{-1}); ρ is the density of water (kg m^{-3}); λ_f, the latent heat of fusion (3.35 $\times$ 10^5 J kg^{-1}); and Q_m, the available energy (W m^{-2}).

To appreciate the significance of a foliage canopy on the magnitude of M it is useful to examine the individual terms in the surface heat balance equation:

$$Q_\mathrm{m} = Q_\mathrm{s}\,(1 - \alpha) + Q_\mathrm{lw} + Q_\mathrm{h} + Q_\mathrm{e} + Q_\mathrm{p} \qquad\qquad [2.44]$$

in which Q_s is the incoming short-wave radiation; α, the short-wave albedo (see table 2.1); Q_lw, the long-wave radiation balance; Q_h, the sensible heat flux; Q_e, the latent heat flux; and Q_p, the heat gained from precipitation.

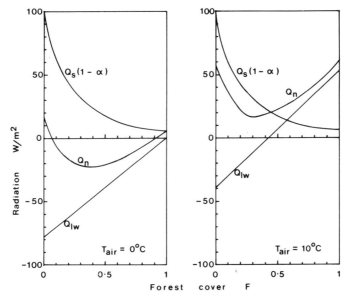

Fig. 2.12 The radiation balance under coniferous forest covers of varying density. Trends correspond to a solar radiation of 242 W m^{-2}, an albedo of 0.6, together with Equation 13-5 (for clear skies) and data from Fig. 13-4 in Dunne and Leopold (1978)

The short-wave radiation — which penetrates the forest canopy and is incident on the snow cover — depends on many factors; solar angle, the type of vegetation, and the amount of foliage on the trees.

Studies conducted by the U.S. Army Corps of Engineers (1956) show that the ratio of Q_s under a forest to that in the open, decreases with increasing foliage density (see fig. 2.12). In contrast to the forested situation, level open snowfields in large openings receive the direct solar beam unimpeded by canopy effects. How much of this direct beam irradiation is absorbed by the snowpack is dependent upon its albedo, which in turn is a function of the age of the uppermost snow layer.

The long-wave radiation received by the snow is comprised of atmospheric long-wave radiation which penetrates the foliage canopy, plus the long-wave input from the canopy itself. These two terms can be weighted in accordance with the amount of foliage cover and are discussed further in U.S. Army Corps of Engineers (1956) and Dunne and Leopold (1978). The melting snow surface may be assumed to have a temperature of 0°C, while the canopy warmed by incoming short-wave and long-wave radiation will have surface temperatures usually equal to, or greater than the ambient air temperature. Net long-wave gain from the forest increases progressively as the canopy cover increases, whereas the net long-wave loss from the snowpack to the atmosphere also decreases progressively (becomes less negative) with increasing canopy cover (see fig. 2.12). The algebraic sum of the long-wave and short-wave streams shows a progressive decrease in the total radiation gain to a canopy closure of about 30% (caused mainly by the reduction in short-wave income) and thereafter the net radiation increases progressively up to a total canopy cover. At this point the net radiation gain is less than that occurring in the open, i.e. canopy density = 0.

The term Q_e refers to the latent heat released to the snowpack when water vapour from the atmosphere condenses on the snow surface. Melt produced by sensible heat transfer or convection results mainly from the heat Q_h transferred from warm air advected over the snow surface. The terms Q_e and Q_h are both dependent on the turbulent transfer processes in the atmosphere near the ground and on other factors depending on the wind speed and the surface roughness. The main effects of vegetation on the terms Q_h and Q_e appear to be linked to the fractions of underlying ground vegetation which protrude through the snow surface and alter the surface roughness; in consequence the turbulent transfer characteristics are altered. Further aspects are discussed in Price and Dunne (1976).

Rain falling on the snow surface at temperatures above 0°C transfers heat to the snow, but this is generally a small contribution. When evaluating snowmelt it is useful to distinguish between periods of rain and rain-free periods.

Snowmelt during rain
For these conditions the predominant heat transfer arises from condensation and sensible heat transfer; heat input from other processes is relatively minor. By adding the various components together the basin snowmelt can be represented as follows:

open and partly forested regions (0-60% cover):
$$M = (1.42 + 0.015 (1 - 0.8F)U_2 + 0.013P)T_2 + 2.29 \qquad [2.45]$$
heavily forested regions (60-100% cover):
$$M = (3.62 + 0.013P)T_2 + 1.27 \qquad [2.46]$$

In these equations each of the melt rates are in mm day^{-1}, T_2 refers to the air temperature 2 m above the snowpack, U_2 (km day^{-1}) is the windspeed (measured at 2 m in the open), P is the mean rainfall in mm day^{-1} and F is the forest canopy cover expressed as a decimal fraction.

Snowmelt during rain-free periods
During rain-free weather, solar and terrestrial radiation (Q_s and Q_{lw} respectively) become the more important melt-producing factors; also the degree of forest cover greatly affects the relative significance of these terms.

The snowmelt equations based on extensive field and laboratory work reported in the U.S. Army Corps of Engineers (1956, 1960) have the following form for a ripe snowpack at 0°C:

Open areas (<10% tree cover):
$$M = 0.258 Q_s (1 - \alpha) + (1 - 0.1C) (1.04T_2 - 21.3) + 0.13CT_c$$
$$+ 0.00078 U_2 (4.2T_2 + 15.1T_d) \qquad [2.47]$$
Lightly forested areas (10-60% cover):
$$M = 0.21 (1 - F) Q_s (1 - \alpha) + 0.00078 U_2 (1 - 0.8F) (4.2T_2$$
$$+ 15.1T_d) + 1.4FT_2 \qquad [2.48]$$
Heavily forested areas (60-80%):
$$M = 0.00078 U_2 (1 - 0.8F) (4.2T_2 + 15.1T_d) + 1.4T_2 \qquad [2.49]$$
Densely forested areas (> 80% cover):
$$M = (1.9T_2 + 1.7T_d) \qquad [2.50]$$

In these equations all the melt rates are in mm day^{-1}; insolation (Q_s) is in W m^{-2}; cloud cover (C) is expressed in tenths of the sky; T_d (°C) is the temperature of the dew-point 2 m above the snow pack; U_2, T_2 and F are as previously defined and α refers to the albedo of the snow pack. T_c (°C) is the temperature of the cloud base which Hendrick *et al.* (1971) suggest can be estimated by the air temperature, say T_2. It should be stressed that equations [2.45] to [2.50] are useful for order-of-magnitude calculations within the stated *ranges* of tree cover and, as stated, are discontinuous at the boundaries of the cover classes. Nevertheless Dunne and Leopold (1978) comment that the equations work surprisingly well in a range of snowy climates (see also Hendrick *et al.* 1971), but add the important caveat that certain of the approximations implicit in equations [2.45] to [2.50] should be checked before embarking on important design studies.

After the melt has been estimated for the portion covered by the snowpack, losses must be deducted, and the melt excess converted into a stream flow hydrograph, by application of unit-graph theory or other methods.

Miller (1959), reviewing both North American and European data, indicated that the melt rate in forests from local heat sources was 75% of melt in the open and suggested this as an approximate general value; this feature demonstrates that the removal of forest results in more rapid melting of snow. In addition, Jeffrey (1970) comments that where the presence of a forest cover results in lowered evapotranspiration, the consequent increase in soil moisture at the end of the summer implies that runoff will be initiated sooner, and will be of greater volume from snow-zone forests following logging. These effects are less pronounced under partial cutting and thinning, than for clear cutting.

Example 2.8 Snowmelt

Compare the likely snowmelt occurring during rainless periods in open and heavily forested areas on the basis of the following data:

$Q_s = 300$ W m^{-2}
$T_2 = 7°C$
$U_2 = 129$ km day^{-1}
$\alpha = 0.6$
$C = 0.4$
relative humidity = 70%

By definition the dew-point temperature T_d is the temperature to which unsaturated air must be cooled to produce a state of saturation, i.e. $e = e_s (T_d)$. At $7°C$, $e_s = 10$ mbar, hence $e = 0.7$, $e_s = 7$ mbar for which $T_d = 2°C$ (see Monteith 1973, page 222).

Open land
From equation [2.47]:
$$M = 0.258 \times 300 (1 - 0.6) + (1 - 0.1 \times 0.4) (1.04 \times 7 - 21.3)$$
$$+ 0.13 \times 0.4 \times 7 + 0.00078 \times 129 (4.2 \times 7 + 15.1 \times 2)$$
$$= 31.0 - 13.5 + 0.4 + 6.0$$
$$= 23.9 \text{ mm day}^{-1}$$

Heavily forested land
From equation [2.49]:
$$M = 0.00078 \times 129 (1 - 0.8 \times 0.7) (4.2 \times 7 + 15.1 \times 2) + 1.4 \times 7$$
$$= 2.6 + 9.8$$
$$= 12.4 \text{ mm day}^{-1}$$
For the conditions stated, this demonstrates a greater rate of melt in the open than under trees and is attributed to the difference in available short-wave radiant input.

2.6 River and channel flows

Flow in open channels is nature's way of conveying water on the surface of the earth through rivers and streams. Whenever there are changes to the land surface and its vegetation, it is likely that these will alter the timing and amount of water flowing – especially peak or flood flows and low flow conditions. Drainage and construction work often require the enlargement or the physical relocation of channels. Conveyance channels are constructed to supply irrigation water or to act as emergency spillways. If insufficient attention is given to the dimensions and form this may result in erosion, deposition and uncontrolled flooding. Although much of hydrology is concerned with transient flow conditions, channels are usually designed or assessed on the basis of uniform flows. The flow equation commonly used for determining the velocity of flow in open channels on the North American continent is the Manning equation, whereas in Europe the Chezy equation is popular. These formulae are empirical, and despite many new proposals for a formula having a theoretical background, the Manning formula still ranks foremost in terms of practical application.

2.6.1 *Manning formula*
The Manning equation is:

$$V = \frac{1}{n} R^{2/3} S^{1/2} \tag{2.51}$$

where V is the average velocity in the cross-section ($m^3\ s^{-1}$); R is the hydraulic radius (m); S, the water surface slope; and n, a roughness coefficient.

The water surface slope can be obtained from slope data, and the hydraulic radius from the ratio $R = A/P$, where A is the cross-sectional area of the flow normal to the direction of flow, and P, the wetted perimeter — which is the length of the line of intersection of the channel-wetted surface, with a cross-sectional plane normal to the direction of flow. The average discharge Q, is calculated by multiplying the cross-sectional area by the velocity determined from the equation. The only difficulty lies in the determination of the roughness coefficient.

The factors which affect Manning's n have been listed and described (Chow, 1959, pages 101-106). These factors may be discussed under three major headings: boundary conditions; conditions of alignment and cross-section; and obstructions. The boundary conditions are defined by the size of bed and bank material; by the type and amount of vegetation on the bed and banks; and on the form resistance created by bed material transport. Manning's n will generally increase with material size, and will increase with the amount of vegetation. The actual effect of vegetation depends to some extent on the flow conditions. For example, the vegetation may be flattened at high flows, producing a low value of n.

Any method of determining n should take account of all the above factors. The only way that this can be accomplished at the present time is by use of tables of n values for various channel conditions (see table 2.21). The use of such tabular data still involves judgement, as a range of n values is given for each channel description. In U.S. Geological Survey (1967) pictures of various channels at different n values represent an aid to the choice of n. Chow (1959) also provides a sequence of pictures that can be used to assist in the choice of the appropriate roughness factor.

Cowan (1956) developed a procedure for estimating n that takes into account several factors which affect the values. The value of n may be computed by his procedure using the equation:

Table 2.21 Values of the roughness coefficient n

Type of channel and description	Minimum	Normal	Maximum
A. Excavated or dredged			
(a) Earth, straight and uniform			
1. Clean, recently completed	0.016	0.018	0.020
2. Clean, after weathering	0.018	0.022	0.025
3. Gravel, uniform section, clean	0.022	0.025	0.030
4. With short grass, few weeds	0.022	0.027	0.033
(b) Earth, winding and sluggish			
1. No vegetation	0.023	0.025	0.030
2. Grass, some weeds	0.025	0.030	0.033
3. Dense weeds or aquatic plants in deep channels	0.030	0.035	0.040
4. Earth bottom and rubble sides	0.028	0.030	0.035
5. Stony bottom and weedy banks	0.025	0.035	0.040
6. Cobble bottom and clean sides	0.030	0.040	0.050
(c) Dragline-excavated or dredged			
1. No vegetation	0.025	0.028	0.033
2. Light brush on banks	0.035	0.050	0.060
(d) Rock cuts			
1. Smooth and uniform	0.025	0.035	0.040
2. Jagged and irregular	0.035	0.040	0.050
(e) Channels not maintained, weeds and brush uncut			
1. Dense weeds, high as flow depth	0.050	0.080	0.120
2. Clean bottom, brush on sides	0.040	0.050	0.080
3. Same, highest stage of flow	0.045	0.070	0.110
4. Dense brush, high stage	0.080	0.100	0.140
B. Natural streams			
B1. Minor streams (top width at flood stage 30 m)			
(a) Stream on plain			
1. Clean, straight, full stage, no rifts or deep pools	0.025	0.030	0.033
2. Same as above, but more stones and weeds	0.030	0.035	0.040
3. Clean, winding, some pools and shoals	0.033	0.040	0.045
4. Same as above, but some weeds and stones	0.035	0.045	0.050
5. Same as above, lower stages, more ineffective slopes and sections	0.040	0.048	0.055
6. Same as 4, but more stones	0.045	0.050	0.060

Table 2.21   cont'd

Type of channel and description	Minimum	Normal	Maximum
7. Sluggish reaches, weedy, deep pools	0.050	0.070	0.080
8. Very weedy reaches, deep pools or floodways with heavy stand of timber and underbrush	0.075	0.100	0.150
(b) Mountain streams, no vegetation in channel, banks usually steep, trees and brush along banks submerged at high stages			
1. Bottom, gravels, cobbles and few boulders	0.030	0.040	0.050
2. Bottom, cobbles with large boulders	0.040	0.050	0.070
B2. Flood plains			
(a) Pasture, no brush			
1. Short grass	0.025	0.030	0.035
2. High grass	0.030	0.035	0.050
(b) Cultivated areas			
1. No crop	0.020	0.030	0.040
2. Mature row crops	0.025	0.035	0.045
3. Mature field crops	0.030	0.040	0.050
(c) Brush			
1. Scattered brush, heavy weeds	0.035	0.040	0.070
2. Light brush and trees, in winter	0.035	0.050	0.060
3. Light brush and trees, in summer	0.040	0.060	0.080
4. Medium to dense brush, in winter	0.045	0.070	0.110
5. Medium to dense brush, in summer	0.070	0.100	0.160
(d) Trees			
1. Dense willows, summer straight	0.110	0.150	0.200
2. Cleared land with tree stumps, no sprouts	0.030	0.040	0.050
3. Same as above, but with heavy growth of sprouts	0.050	0.060	0.080
4. Heavy stand of timber, a few down trees, little undergrowth, flood stage below branches	0.080	0.100	0.120
5. Same as above, but with flood stage reaching branches	0.100	0.120	0.160
B3. Major streams (top width at flood stage 30 m); the n value is less than that for minor streams of similar description, because banks offer less effective resistance			
(a) Regular section with no boulders or brush	0.025	–	0.060
(b) Irregular and rough section	0.035	–	0.100

$$n = (n_0 + n_1 + n_2 + n_3 + n_4)\, m_5 \qquad\qquad [2.52]$$

where

n = value to be used

n_0 = a basic n value for a straight, uniform smooth channel in the natural materials involved

n_1 = a value to correct for the effect of surface irregularities

n_2 = correction factor for variations in size and shape of the channel

n_3 = correction factor for the effects of obstruction

n_4 = correction factor for the vegetation and flow conditions

m_5 = correction factor for the meandering of the channel

Table 2.22 Values for the computation of the roughness coefficient

	Channel conditions		Values
Material involved	Earth	n_0	0.020
	Rock cut		0.025
	Fine gravel		0.024
	Coarse gravel		0.028
Degree of irregularity	Smooth	n_1	0.000
	Minor		0.005
	Moderate		0.010
	Severe		0.020
Variations of channel cross-section	Gradual	n_2	0.000
	Alternating occasionally		0.005
	Alternating frequently		0.010 – 0.015
Relative effect of obstructions	Negligible	n_3	0.000
	Minor		0.010 – 0.015
	Appreciable		0.020 – 0.030
	Severe		0.040 – 0.060
Vegetation	Low	n_4	0.005 – 0.010
	Medium		0.010 – 0.025
	High		0.025 – 0.050
	Very high		0.050 – 0.100
Degree of meandering	Minor	m_5	1.000
	Appreciable		1.150
	Severe		1.300

Values of n and m may be selected from table 2.22. The conditions applying for each value of n and m are explained in Chow (1959, pages 106-108). It should be stressed that this method is limited to conditions in which no sediment transport occurs and further, that the values were derived for small channels $R < 4.5$ m, so they should be used with caution for larger channels.

When considering the effects of vegetation, it should be recognised that owing to the seasonal growth of aquatic plants, grass, weeds, shrubs, and trees in the channel or on the banks, the value of n may increase in the growing season and diminish in the dormant season (see Miers, 1974). This seasonal change may cause changes in other factors.

When planning river works, the Institution of Water Engineers (1969, page 296) recommends that it is preferable to be guided by the measured value of n applying to the channel before improvement, unless improvement is very drastic.

2.6.2 Flow in a channel section with composite roughness

In natural streams it is often the case that the bottom of the stream is free of vegetation whereas the banks are covered. As the discharge increases and the river level rises, the banks become submerged; here, one faces the situation of the considerable variation in the roughness along the wetted perimeter. In applying Manning's equation to such channels it is sometimes necessary to compute an equivalent n value for the entire perimeter and to use this equivalent value for the computation of the flow in the whole section.

Various techniques are available for assessing the equivalent n value (see Chow, 1959, pages 136-40). For example, if the water area is divided into N parts of which the wetted perimeter $P_1, P_2, \ldots P_N$ and the coefficients of roughness $n_1, n_2, \ldots n_N$ are known, then with the assumption that each part of the area has the same mean velocity (= mean velocity of the whole section) then:

$$\overline{n} = \frac{\sum\limits_{i=1}^{N} P_i N_i^{1.5}}{P} \qquad [2.53]$$

Another situation which is often encountered in flood control is the provision of flood berms on either side of the main channel, to provide storage during extreme high flood discharges (fig. 2.13). Flood berms are often used for grazing and other agricultural purposes; they are generally rougher than the main channel. In river management, the main channel is usually designed to carry normal winter discharges, whereas the exceptional flows spill into the flood plains on either side. For channels of compound section, the Manning formula may be applied to each subsection in determining the mean velocity of the subsection; then the discharge in the subsection can be computed. The total discharge is the sum of these discharges and the mean velocity of the channel section is equal to the total discharge divided by the total water area (see Chow, 1959, pages 138-40 and example 2.9).

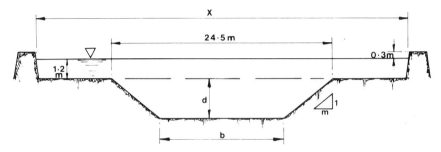

Fig. 2.13 Channel cross-section with flood berms

Example 2.9 Channel design

The cross-section of a river channel shown in fig. 2.13 is to be designed in such a way that normal winter discharges are carried in a main central channel while extreme high flows spill over the flood berms on either side bounded by low banks constructed parallel with the river. The following particulars apply:

(a) The crest of the flood bank is 1.5 m above the flood plain and allows a free board of 0.3 m to cater for wave action.
(b) The main channel section is trapezoidal with side slope 1:2 and overall width 24.5.m.
(c) The longitudinal slope of the river channel is 1:1000.
(d) $n = 0.04$ for main channel and 0.05 for flood plains, which are covered with scattered brush and tall grass.
(e) The main channel section is designed to pass a bank-full discharge (Q_0) of 82 m³ s⁻¹.
(f) Flood banks are sited so as to pass a total peak flood discharge (Q_p) of 210 m³ s⁻¹.

Determine the depth and base width of the main channel and the minimum distance between the flood banks.

Main channel – bank full condition ($Q_0 = 82$ m^3 s^{-1})
For a trapezoidal channel of base width b, depth d and side slope
$1:m$ ($m = 2$):
Area of channel, $A_1 = 24.5\, d - md^2 = 24.5d - 2d^2$ [2.54]
Wetted perimeter, $P_1 = b + 2d\,(1 + m^2)^{1/2} = (24.5 - 4d) + 2d\sqrt{5}$
        [2.55]
Hence, $R_1 = A_1/P_1 = d(24.5 - 2d)/(24.5 + 0.47d)$ [2.56]
The flow is defined by the product: cross-sectional area X channel
velocity (from Manning's equation 2.51):

$$Q_0 = A_1\, R_1^{2/3} \times (\frac{1}{1000})^{1/2}/0.04 \qquad\qquad [2.57]$$

which is satisfied by $d = 3.35$ m. Thus $A_1 = 59.6$ m^2, $P_1 = 26.1$ m and
$b = 24.5 - 4d = 11.1$ m.

Peak flow condition ($Q_p = 210$ m^3 s^{-1})
For the purpose of the evaluation, the cross-section is divided into a
main channel and a shallow side channel, the latter representing the
flood plain. These channels are regarded as operating independently
with a differing surface roughness and mean velocity.

(a)   Main channel:
     $A_2 = A_1 + (1.5 - 0.3) \times 24.5 = 89.0$ m^2 [2.58]
     Since the wetted perimeter is unaltered:
     $R_2 = A_2/P_1 = 3.41$ m   [2.59]
     Following the logic of equation [2.57] yields:
     $Q_2 = A_2\, R_2^{2/3} \times (\frac{1}{1000})^{1/2}/0.04 = 159$ m^3 s^{-1} [2.60]

(b)   Side channel:
The required capacity of the flood plain is $Q_p - Q_2 = 51$ m^3 s^{-1}. As
the section is wide and shallow, $R =$ depth of flow (1.2 m). Hence:

$$\underbrace{51}_{\text{discharge}} = \underbrace{1.2\,(x - 24.5)}_{\text{area}} \times \underbrace{1.2^{2/3} \times (\frac{1}{1000})^{1/2}/0.05}_{\text{velocity}}$$

giving $x = 84$ m as the minimum distance between the flood banks.

2.7 Hydrology and land use

Whereas previous discussion has focused on the role of vegetation within the principal hydrologic processes occurring at the land surface, here we turn our attention to their role in field situations. These provide an array of problems which vary with scale, land use (urban, agricultural, forest, etc.) and the prevailing climate (wet, arid, snowy, etc.). Some problems focus on primary variables, such as water yield and floods, whereas others include secondary effects, such as erosion, water quality and salination. Each problem presents unique features dependent on local circumstances and clearly it is not feasible to discuss all eventualities. Despite their differences, however, all conform to basic laws — of which mass conservation and energy conservation are particularly important.

2.7.1 Agricultural lands and small catchments

The successful development of agricultural land is inextricably linked with water management. Although practices extend into upland areas, we are mainly concerned with rural lowland areas. Downstream, it is the fertile flood plains which form much of the prime agricultural land. Cities, roads and railways are built to take advantage of level ground together with the waterway's usefulness for water supply, sewage disposal and navigation; as a result floods are the most damaging of all natural catastrophes. Most floods can be 'controlled' by the use of engineering structures and careful management of upstream land and vegetation. Crop growers are concerned primarily, and often exclusively, with the control of water reaching the land — in order that it will do the least harm and be of greatest benefit to the crops produced. Where there is insufficient rainfall this may focus on the problems of irrigation. Alternatively, where the soil is underlain with material of low permeability, or where the water table is high, plant growth may be impaired; the crop grower must turn his attention to drainage in order to improve the agricultural potential. Land devoted to cultivated crops is a form of agricultural land use which is least receptive to rainfall — especially during periods when the plant cover is small and the soil exposed. At this most vulnerable stage, runoff carries the threat of erosion — which may skim fields of their most productive soil, and make the cultivation of land difficult.

Many problems centre on the water balance since it is the interrelationships between inputs, outputs and storage which ultimately determine the disposition of water at the land surface. This may embrace the entire catchment or focus on particular soil profiles as in crop irrigation. In the following section a number of hydrologic situations are examined to provide a perspective on the role of particular hydrologic processes and of their interaction in the agricultural domain.

2.7.1.1 ANNUAL WATER BALANCE OF AGRICULTURAL CATCHMENTS

The annual water balance of a catchment can be established by postulating that the mean annual evaporation must equal the difference between mean precipitation and mean runoff as depicted in equation [2.2]. Penman (1951) showed that the values $P - R$ for many British catchments were consistent with estimates of potential evaporation derived from climatological records. When evaporation is determined on a monthly basis, seasonal changes in the amount of water stored within an impermeable catchment can be estimated from differences between the monthly income of water and its loss by evaporation and stream flow. In principle, evaporation (including interception) can be estimated using the Penman formula for short vegetation. In practice, however, one must account for the availability of water and its interaction with growth characteristics of the surface vegetation. Simple relationships between transpiration and soil water are difficult to establish and the performance of a catchment is more complex. This aspect was clearly demonstrated by McGowan *et al.* (1980) in an analysis of the water balance representative of much of eastern and central Britain. Their data conformed to a pattern in which the annual evaporation, calculated as $P - R$, followed the general relationship:

$$P - R = a + bP \qquad [2.61]$$

with a and b as 'constants'; this suggests that annual evaporation is limited by rainfall; it also shows that catchment yield increases linearly with precipitation, a feature which has been observed elsewhere (see Gray, 1970, section 10).

By distinguishing between 'winter' evaporation (E_w) and 'summer' evaporation E_s, $P - R = E_w + E_s$. McGowan *et al.* (1980) found that E_w was consistent with the Penman estimate E_T, whereas during the summer months $E_s < E_T$ and conformed to the pattern:

$$E_s = c + fP_s \qquad [2.62]$$

where c represented the maximum amount of water stored in the soil and available for evaporation, and f is a fraction of P_s, the precipitation over the same period. Thus it is seen that the catchment behaves like a system in which the summer evaporation is usually determined by the amount of precipitation and by the reserve of water in the soil, i.e. *it is independent of the potential rate of evaporation determined by the weather*.

Geological factors aside, the area conforming to this pattern of behaviour can be roughly assessed by determining the amount of summer rainfall (P_s') required to reach the annual equivalent of potential evaporation:

$$P_s' = [E_T \text{ annual} - (c + E_w)]/f \qquad \text{(from equation [2.62])}$$

and comparing this with the pattern of summer rainfall. If $P_s < P_s'$, then it is likely that summer rainfall restricts evaporation and runoff; whereas if $P_s > P_s'$, then the annual water balance tends toward $P - R = E_T$, with evaporation taking place at the potential rate. Gray (1970) suggests that linear annual precipitation-runoff relationships may be appropriate where the precipitation is moderate and well distributed – such as in temperate, sub-humid regions.

McGowan et al. (1980) also showed that evaporation rates departed from the potential rate when the soil moisture deficit exceeded 40 mm. The figure 40 mm represents a mean 'root' constant for the catchment, and is less than the value adopted in Penman (1949).

An additional factor affecting evaporation from an agricultural catchment is the type of vegetative cover. If we neglect the particular problems posed by trees, it is the differences in annual water use between pastures and cereal crops which have attracted most attention in recent years. Differences in annual water demand may be attributed to a number of factors such as the extent of ground cover, rooting depths, the response of transpiration to soil moisture and the relative length of the growing season. Generally, the annual water loss by pasture exceeds that of cereals and implies that if pasture areas were replaced by cereals within a catchment, this should lead to an increase in runoff (McGowan et al. 1980).

2.7.1.2 STORM RUNOFF

Besides requiring knowledge of annual trends in runoff, engineers are frequently concerned with control of water from selected storms. Runoff volume can be estimated directly from the surface water balance, paying particular attention to the soil moisture status and the infiltration character. Alternatively, it may be inferred from schemes such as those described in section 2.4.2 which takes account of soil type, soil moisture status, precipitation rate and land use. The Rational method (section 2.4.3) is useful for design (as distinct from prediction) in small catchments, and in rural areas is usually concerned with storms with a 5-10 year frequency (see table 2.23). The Rational runoff coefficient makes some provision for particular forms of land use, but greater detail appropriate to agricultural catchments is described in Schwab *et al.* (1966). The unit hydrograph (section 2.4.4.1) represents a more advanced system for identifying flood character and has much merit when applied to small watersheds. For more sophisticated analysis, design can be based on the mean annual flood using the techniques described in Natural Environment Research Council (1975), or simulation techniques, e.g. Holtan *et al.* (1975).

Table 2.23   Flood return period for design purposes

Type of agriculture	March to mid-Nov. (years)	Minimum acceptable standard over the whole year (years)
Horticultural crops including those grown on field scale	100	100
Root crops, cereals and intensive grazing	25	10
Cereals and intensive grass only	10	5
Grass	5	2
Extensive grazing	3	1

2.7.1.3 CROP IRRIGATION

Crop irrigation is an age-old art which continues to be practised on an ever-increasing scale. In 1975 it was estimated that 233 million hectares were under irrigation – representing about 16% of the world's cultivated land (Hansen *et al.* 1980). Though irrigation focuses on crop production and the application of water to the soil, the scope of irrigation science extends from the catchment to the farm, and on to the drainage channel. Catchment features affect the yield of irrigation water; they also have a major bearing on the degree of silting within storage dams and conveyance channels. Water losses during conveyance are particularly important; these may be due to seepage from unlined channels or perhaps caused by phreatophytes – water-loving plants growing along stream courses or on wet soils where water is readily available. Phreatophytic vegetation has little or no economic value and consumes water which would otherwise be available for the irrigation of crops. The problems posed by this type of vegetation are particularly acute in arid areas and in areas possessing a high water table. The design of surface and subsurface drainage systems is also important; proper consideration should be given to the disposition of excess water. Finally, in every river basin, prior to the introduction of irrigation, there exists a water balance between the rainfall and stream flow, groundwater and evaporation. This balance is disturbed when irrigation water is added. For example, seepage losses during conveyancing, or water supplied in excess of the requirements of plants, can make a significant contribution to groundwater and may alter the drainage of the area. Irrigated agriculture also changes the distribution of salts in the soil profile; there are many large areas in the Middle East and South-east Asia which have become saline or alkaline and unproductive.

To establish an irrigation programme, one must have adequate information on water supply, climate, plants, soil and economic factors. In order to determine whether there is sufficient water, estimates must be made of the total water requirements, including evapotranspiration losses, runoff, deep percolation, conveyance losses and necessary leaching. In the long-term, precautionary measures should be taken to prevent the long-term damage of irrigation projects, e.g. arising from salination, whereas in the short-term, there is a general emphasis on optimising water supply – avoiding shortages and excesses. Limited water supply leads to growth retardation, whereas excesses produce heavy percolation. If the water table is raised, this may restrict the space available for good root development. Irrigation water supply needs to be matched to the crop water requirement – this being a function of climate, crop and the particular growth stage. In withdrawing water from the soil, plants are able to exclude the bulk of salts in the soil solution. Thus regular supply of salts-containing water to the root zone results in a build up of salts in the soil (which may damage its structure and impair growth) unless they are leached out by excess water.

Since the main cause of failure of irrigation projects is the possibility of waterlogging and salination, it is these aspects which pose the main technical problems in determining the quantity of water which must be applied. In the first instance, this can be evaluated by maintaining acceptable water and salt balances within the whole drainage area, or within sub-areas. Salt balances are similar to water balances in that they describe gains and losses of a salt in a given area or soil layer over a certain period of time (see section 4.8). Here we shall focus on the water balance, but recognise that the approach must be modified when salinity problems are present.

Crop water requirements
Various methods exist for assessing crop water requirements and the following outlines the scheme described in Doorenbos and Pruitt (1977): they defined the crop water requirements as, 'the depth of water needed to meet the water loss through evapotranspiration (E_c) of a disease-free crop, growing in large fields under non-restricting soil conditions, including soil water and fertility, and achieving full production potential under a given growing environment'. To calculate E_c a three stage procedure is recommended:

(a)   The effect of climate on crop water requirements is given in terms of a reference crop evapotranspiration and is consistent with the definition of E_T (equation [2.12]) with the reference crop as short grass. E_T may be estimated using the principles described in section 2.2.3 or by other techniques such as the Blaney-Criddle method (see U.S. Soil Conservation Service, 1970).
(b)   The effect of crop characteristics on crop water requirements is based on the relationship:

$$E_c = k_c E_T \qquad\qquad\qquad [2.63]$$

in which k_c is a crop coefficient, which depends on the type of crop, the stage of growth, the growing conditions and the prevailing weather conditions.
(c)   The effect of local conditions and agricultural practices on crop water requirements.

Selection of crop coefficient
For field and vegetable crops the growing season may be divided into four stages as illustrated in fig. 2.14. In the initial stage (ground cover <10%), the water loss is mainly by evaporation from the base soil surface, because in the absence of an established root system, water cannot easily be raised from the lower layers. If the soil surface is wet, the evaporation rate is almost that of a free water surface, but

falls rapidly as the surface dries. Thus a dominating factor influencing k_c, is the number of days since the soil became wet by rainfall or irrigation. The value indicated in fig. 2.14 is a smoothed trend commensurate with the water requirements. During the crop development stage, water loss by transpiration from the developing leaves becomes increasingly important until the ground cover reaches 70-80%, at which evapotranspiration attains the potential rate. This rate is maintained during the mid-season stage while the crop matures; for most crops $k_c \doteqdot 1$, i.e. is essentially independent of crop type. Finally, during the late season phase as the crop approaches maturity, the surface resistance increases and k_c is reduced. Doorenbos and Pruitt (1977) provide a summary of k_c as a function of crop type, growth stage, wind speed and relative humidity.

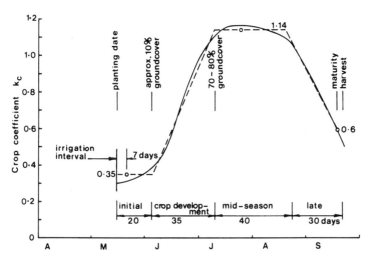

Fig. 2.14 Example of crop coefficient curve

The total crop water requirement can be assessed from the summation of the monthly estimates of E_c together with small correction factors. From this, the net irrigation requirement (I_n) is determined from the approximate water balance

$$I_n = E_c - P_e \qquad\qquad [2.64]$$

where P_e is the effective rainfall; this refers to the amount of rainfall available to the crop during the growing season and is defined by:

$$P_e = P_g - (R + E_n) \qquad\qquad [2.65]$$

where E_n is the evaporation from the soil surface in the non-growing season and other parameters as previously defined. Where salinity problems exist, additional water will be required for leaching accumulated salts from the root zone, as described in section 4.8.

Daily demand

Day-to-day management of irrigation schemes is often concerned with the control of the soil moisture deficit. Most crops can tolerate a soil moisture deficit before there is a significant effect on growth; the maximum acceptable values of *SMD* depend on the crop, the stage of growth and the soil. Typical values for crops grown in Britain are summarised in table 2.4. The soil moisture status is assessed using:

$$\Delta SMD = E_c - P_g - I_i \qquad\qquad [2.66]$$

where I_i is the input by irrigation. E_c depends on the selected value of k_c and an evaluation of E_T. In the United Kingdom the Meteorological Office provides estimates of E_T within specified grid squares over periods of seven days. For planning purposes, estimates of E_T may be obtained in advance using average data such as in Ministry of Agriculture, Fisheries and Food (1967). Correction should be made for deviations between the actual and average estimates of E_T. While the E_T data can be supplied from central sources, P_g must be measured locally, because rainfall varies too much over short distances. By following the soil moisture deficit using a daily water balance, irrigation can be initiated each time the value of *SMD* reaches the maximum allowable level, and it is common practice to restore the soil moisture to just below field capacity (see fig. 2.15). It is usually agreed that production is maximum if less than 50% of the available water is removed during the vegetative, flowering and wet fruit stages; adjustments are made to satisfy the particular needs of crops.

It should be stressed that the water requirements predicted by equations [2.57] and [2.58] are the target water needs. Actual water requirements must be adjusted to take account of the inefficiencies in distribution and application.

For further discussion the reader is referred to Hansen *et al.* (1980), Dastane (1974) and Ministry of Agriculture, Fisheries and Food (1982) and Food and Agriculture Organisation (1973), these representing a small sample of an extensive literature on irrigation principles, techniques and practices.

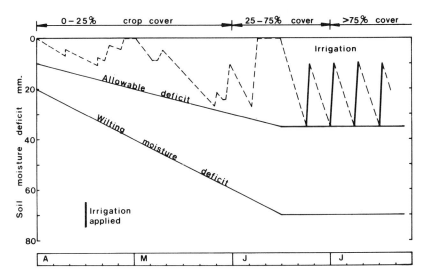

Fig. 2.15 Soil moisture changes with variable rooting depth

2.7.1.4 TRANSPIRATION AND DRAINAGE CONTROL

A problem of a different nature and dependent on the surface water balance, concerns the production of leachates from sanitary landfills and mineral spoil tips. By infiltrating and seeping through the waste media, rainwater (reappearing as drainage water) may be grossly polluted — threatening the quality of receiving waters. Several costly methods are available for resolving this problem. One approach is to cap the tip with an impervious layer to prevent infiltration. Otherwise the leachate may be collected and be treated. An alternative is to reduce the amount of drainage water by exploiting transpiration as a mode of water loss.

Inspection of the water balance (section 2.1.2) indicates that the amount of drainage water over a lengthy period of time is given by:

$$Q = P_g - E \qquad\qquad [2.67]$$

Here Q represents the maximum quantity of water available for infiltration which will be reduced if surface runoff takes place. For present purposes we shall ignore this distinction and consider Q as draining through the tip and re-emerging as a delayed runoff in the form of leachate.

Actual evapotranspiration E consists of evaporation from the soil and transpiration, and it is necessary to separate these components when evaluating the water balance over a surface partially covered by vegetation. Evaporation from bare soil is subject to the same controls as evaporation from an open water surface, i.e. it takes place at about the potential rate when the surface is wet. However, if the surface becomes dry, there is a rapid drop in evaporation because vapour must diffuse upward through stagnant air in the soil pores and this is a slow process. A simple model employed in agricultural practice in the United Kingdom (see Grindley, 1969) is to assume that evaporation takes place at the potential rate until 25 mm of moisture has been extracted; evaporation then ceases until more rain is added.

Table 2.24   The dependence of drainage yield on cover type

Month	P_g (mm)	E_T (mm)	$P_g - E_T$	C = 200 mm Deficit	Drainage	C = 75 mm Deficit	Drainage	Bare soil Deficit	Drainage
March*									
April	23	50	−27	27		27		25	
May	51	81	−30	57		57		25	
June	12	95	−83	140		111		25	
July	72	100	−28	168		145		25	
Aug	48	76	−28	196		116		25	
Sept	52	53	−1	197		116		25	
Oct	109	29	+80	117		36			55
Nov	65	8	+57	60			21		57
Dec	66	7	+59	1			59		59
Jan	106	11	+95		94		95		95
Feb	22	18	+4		4		4		4
March	16	26	−10	10		10		10	
Σ	642	554			98		179		270

*Assume that there is no deficit at the end of the month

The implications of this pattern of behaviour are shown in table 2.24 which has been constructed to illustrate the effect of long-rooted vegetation (trees, etc.), short-rooted vegetation (grass, etc.) and bare soil on the yield of drainage water. Long-rooted vegetation is assumed to be able to draw freely on 200 mm of soil moisture whereas the root constant (C) for short-rooted vegetation is taken as 75 mm. The method of calculation follows the procedures discussed in section 2.2.6. As might be expected the yield is least from deep-rooted vegetation and most from bare soil, a feature which was observed by Molz and Browning (1977) in experiments with lysimeters. The differences in yield depend on the response of actual evaporation to the soil moisture deficit; had there been more rainfall in the summer months shown in table 2.24 the resultant yields would converge.

Overall, the example suggests that if a waste tip is located in a dryish climate then leachate volumes can be reduced by maintaining a vegetative cover. The benefit can be roughly assessed by obtaining weekly or monthly climatological data and following the procedures shown in table 2.24. Thereafter, it is a matter of weighting the yields according to the fractional areas covered with vegetation and bare soil.

2.7.2 Forest lands

Forests are not only an important economic resource yielding wood products, but also produce benefits of flood mitigation, erosion control and landslide prevention. In addition, they offer recreational uses and conservation values − aspects which are becoming increasingly cherished within the industrialised world. Forestry operations now penetrate to realms of catchments previously immune from the power of the bulldozer; as prospecting techniques improve and the economic climate changes, mining operations are being introduced into the more remote upland areas to exploit deposits. Agricultural land area is continually being expanded at the expense of forests; one also witnesses the increased use of fertilisers and biocides to sustain yields. All these produce effects which are influential on hydrology, and are particularly prevalent where forests are situated in small watersheds.

As a general introduction to forest hydrology the reader is referred to Lee (1980), Gray (1970) and Storey *et al.* (1964), these providing useful insight into basic principles and the hydrologic assessment of forestry operations.

2.7.2.1 WATER YIELD

There is considerable evidence, e.g. Shachori and Michaeli (1965), Hibbert (1967) and Clarke and Newson (1978) to indicate that the water yield from forested catchments is less than from comparable areas covered by pasture or short vegetation and this is caused by the following factors. First, forests have a lower albedo (table 2.1) and this implies that there is more available energy to support evaporation. Second, the increased aerodynamic roughness of a forest magnifies the effective ventilation component of evaporation (see equation [2.12]). Third, in wetter climates interception losses are greater than from short crops (section 2.2.7). Finally, the rooting depth of trees may have greater access to available water and continue to transpire at a normal rate during conditions in which short-rooted vegetation would be forced to reduce transpiration rates (section 2.2.5).

Manipulation of vegetation for water yield

As water shortages become more severe attempts are being made to increase the water yield from drainage basins by manipulating vegetation (e.g. Storey and Reigner, 1970 and Hibbert, 1981). In forests this is achieved by complete removal (clear felling or clear cutting), replacement of one tree species by another or by structural alterations in existing stands.

After clear felling there will be a large and immediate increase in runoff caused by the marked reduction in transpiration; Galbraith (1975) and Lee (1980) describe procedures for predicting the increase in water yield. Generally speaking, increases in water yield are greatest in the first year after felling. Thereafter, the general pattern is for the yield to decline as vegetation becomes re-established.

When one species of vegetation is replaced by another it is difficult to specify the effect without knowledge of the local climate or the species involved. Encouragement should be given to deciduous trees, since these will intercept less in winter and have a shorter season of transpiration – thus using less water. The significance of climate was discussed in Calder (1979), whose review indicated that the water loss from a forest in a drier region in the United Kingdom was less than the comparable loss that would have occurred from grassland. However, the situation is reversed in wetter climates due to the increased significance of interception. First-order estimates of the role of interception may be obtained from using equation [2.26]. This was used by Calder and Newson (1979) to assess the anticipated reduction in water yield resulting from plans for extensive afforestation in the uplands of Britain.

Another procedure to increase water yield from forest land involves snowpack management. This involves cutting in strips to redistribute heavy snowfalls and to reduce evapotranspiration. Shading of the snow by the remaining trees is also a factor saving water. The step-and-wall system of strip cutting suggested by Anderson (1963) from his studies in the High Sierras (California) is an example of this approach. Strips are oriented and cut so that the uncut forest shades the open area as much as possible. Similar findings have been recorded in the Fool Creek watershed near Fraser in the Rocky Mountains of California. Here, it was demonstrated that snow redistribution did not lead to a measurable increase in total snow storage in a treated area compared with an untreated control; yet the treatment increased water yield by a significant percentage (Hoover, 1969).

Example 2.10 Afforestation and water yield

Plans exist to increase the proportion of forested area from 11% to 84% in an upland catchment in Britain, the remaining land being pasture. Given that the annual potential evaporation from the area is 518 mm and the annual precipitation 2348 mm, with average rainfall intensity of 1.4 mm h^{-1}, estimate the potential reduction in runoff caused by interception, assuming that 30% of the annual precipitation incident on the forest is lost by interception.

From equation [2.27]:
$$\text{Fraction of year when canopy is wet } (w) = \frac{2348}{1.4} \times 1.71 \times 10^{-4}$$
$$= 0.29$$

Before afforestation:
Annual evaporation* = 518 + 0.11 × [2348 × 0.3 − 0.29 × 518]
= 579 mm
Annual runoff = 2348 − 579
= 1769 mm

After afforestation:
Annual evaporation* = 518 + 0.84 × [2348 × 0.3 − 0.29 × 518]
= 984 mm
Annual runoff = 2348 − 984
= 1364 mm

Hence the runoff associated with rain is reduced by 23%.

2.7.2.2 DIRECT RUNOFF

One of the most important functions of forests is their effect in regulating runoff; the reasons are complex and interrelated, but may be largely attributed to the excellent infiltration characteristics of well-established forest soils. Because of this tendency to recharge the ground reservoir, it permits the slow release of water as baseflow to streams. With a good rooting depth forests have access to deeper moisture supplies than are available to shorter vegetation; this is valuable for supporting transpiration during dry periods and allows the establishment of substantial soil moisture deficits – an important buffer against surface runoff. Interception by the canopy and surface litter reduces the volume available for runoff, and also introduces a time delay which facilitates infiltration. An extensive forest cover

*From equation [2.26] and subject to the stated conditions.

also delays snowmelt and allows a slower release of water than would otherwise occur in the open (see section 2.5.2). By decreasing surface runoff, the erosion rate under forest is considerably less than that under other land uses (see section 3.4.1). Low erosion rates also help to maintain the infiltration capacity of the soil by reducing the possibility of clogging of soil pores with fine particles.

From this brief discussion it is apparent that any major disturbance such as clear cutting and forest fires, will seriously affect these qualities and often result in more frequent floods and greater erosion. As an example consider some of the effects of logging operations. Here the removal of vegetal cover causes a temporary reduction in evapotranspiration and hence increases the water yield; it implies that the soil moisture in logging areas is increased so that the final infiltration capacity (v_c) is reached more quickly than in uncut areas; the loss of the canopy also increases the net precipitation. An extensive use of vehicles during logging will cause soil compaction — reducing its permeability and infiltration capability. Many of these features are also likely to arise following forest fires, but fires are likely to promote a more radical disturbance to forest soils than is caused by timber extraction (Jeffrey, 1970).

Flow extremes
In regions where forests occur, total discharge, stormflow volume and peak discharge are generally smaller than from adjacent non-forested land. When discussing the effects of forests on peak flows, care should be taken to distinguish between rain and snow, the moisture status of the catchment, and the amount of precipitation. When floods emerge from a completely saturated land, e.g. after prolonged and heavy rainfall, land use has relatively little effect on the stream flow. On the other hand, if floods result from the rainfall intensity exceeding the infiltration capacity, then land use has a more significant impact on the resultant discharge. In this respect forests are a more convenient buffer than other forms of land use. A further aspect is the situation of floods released from snowmelt; here, forest cutting can increase the amount of radiation reaching the snowpack and may result in accelerated runoff.

Because of the great variety of conditions encountered, it is difficult to generalise about the precise effects of land treatment on peak flows, beyond the observations which have already been made. In practice, use is often made of multiple regression analysis, e.g. Storey *et al.* (1964). This approach can relate the peak discharge to a series of climatic and catchment parameters including the cover density;

good correlations emerge and the empiricism establishes a hydrologic base against which potential land treatment effects can be compared.

From the point of view of flood mitigation the ideal is to maintain a full and dense canopy allowing maximum interception. Preference should be given to deep-rooted evergreens, since these are likely to promote the deepest soil drying. Steps should always be taken to maintain a fast-growing healthy stand, protecting the forest cover from fire, insect damage, and uncontrolled wood-cutting and grazing. As an economic resource it is essential that logging operations are able to continue; due consideration must be given to the physical characteristics of the watershed as well as to the local climate. In Japan, where tracts of forest are retained for flood protection, forest cutting is carefully controlled. Cuts are made in blocks and the total area allowed for an annual cut is determined according to the formula:

$$A = \frac{F}{U} \qquad\qquad\qquad\qquad [2.68]$$

where A is the allowable area that can be cut per year; F, the total area of the protection forest where clear cutting is permitted; and U is the predetermined rotation age (or cutting age). For example, if the total area to be cut is 35 000 ha and the final cutting age is 40 years, then the annual cutting area is 875 ha. Further restrictions and related aspects are discussed in Forestry Agency, Japan (1976).

The influence of forest cutting on low flows is of great practical importance because the increases in flow occur when the stream flow is in greatest demand. Average low-flow increases must be of the same order as total yield increases and seasonal variations follow the same trends. In some cases afforestation may result in permanent streams becoming ephemeral.

2.8 References

American Society of Civil Engineers (1949) *Hydrology Handbook*. A.S.C.E. Manuals of Engineering Practice No. 28. New York: American Society of Civil Engineers.

Anderson, H.W. (1963) 'Managing California's snow zone lands for water'. Research Paper PSW-6, Pacific Southwest Forest and Range Experiment Station: U.S. Department of Agriculture, Forest Service.

Baumgartner, A. (1972) 'Water and energy balances of different vegetation

covers'. *Proceedings of the Reading Symposium on the World Water Balance*
3 (July 1970), 570-80. Published jointly by International Association of
Scientific Hydrology (Belgium), World Meteorological Organisation (Geneva)
and Unesco (Paris).

Bell, F.C. and Songthara om Kar (1969) 'Characteristic response times in design
flood estimation'. *Journal of Hydrology* **8**, 173-96.

Blake, G.J. (1975) 'The interception process'. In Chapman, T.G. and Dunin,
F.X. eds. *Prediction in Catchment Hydrology*. Canberra: Australian Academy
of Science, 59-81.

Body, D.N. (1975) 'Empirical methods and approximations in the determination
of catchment response'. In Chapman, T.G. and Dunin, F.X. eds. *Prediction in
Catchment Hydrology*. Canberra: Australian Academy of Science, 293-304.

British Standard CP 2005 (1968) *Code of Practice for Sewerage*. London: British
Standards Institution.

Calder, I.R. (1977) 'A model of transpiration and interception loss from a spruce
forest in Plynlimon'. *Journal of Hydrology* **33**, 247-65.

Calder, I.R. (1979) 'Do trees use more water than grass?' *Water Services (G.B.)*
83, 11-14.

Calder, I.R. and Newson, M.D. (1979) 'Land-use and upland water resources in
Britain — a strategic look'. *Water Resources Bulletin* **15**, 1628-39.

Chow, V.T. (1959) *Open-Channel Hydraulics*. New York: McGraw-Hill.

Chow, V.T. (1964) *Handbook of Applied Hydrology*. New York: McGraw-Hill.

Clarke, R.T. and Newson, M.D. (1978) 'Some detailed water balance studies on
research catchments'. *Proceedings of the Royal Society of London*. Series A,
363, 21-42.

Cowan, W.L. (1956) 'Estimating hydraulic roughness coefficients'. *Agricultural
Engineering* **37**, 473-75.

Crawford, N.H. and Linsley, R.K. (1966) *Digital Simulation in Hydrology:
Stanford Watershed Model IV*. Technical Report No. 39. Stanford, California:
Stanford University.

Dastane, N.G. (1974) *Effective Rainfall in Irrigated Agriculture*. F.A.O. Irrigation
and Drainage Paper No. 25. Rome: Food and Agriculture Organisation of the
United Nations.

Delfs, J. (1955) *Die Niederschlagszüruckhaltung im Walde*. Mitteilungen des
Arbeitskreises 'Wald und Wasser'. Koblenz.

Doorenbos, J. and Pruitt, W.O. (1977) *Guidelines for Predicting Crop Water
Requirements*. F.A.O. Irrigation and Drainage Paper No. 24. Rome: Food and
Agriculture Organisation of the United Nations.

Dunne, T. and Leopold, L.B. (1978) *Water in Environmental Planning*. San
Francisco: Freeman.

Fleming, G. (1975) *Computer Simulation Techniques in Hydrology*. New York:
Elsevier.

Fleming, P.M. and Smiles, D.E. (1975) 'Infiltration of water into soil'. In
Chapman, T.G. and Dunin, F.X. eds. *Prediction in Catchment Hydrology*.
Canberra: Australian Academy of Science, 83-110.

Food and Agriculture Organisation (1973) *Irrigation, Drainage and Salinity*. Kovda, V.A. *coordinating ed*. F.A.O. and Unesco. London: Hutchinson.

Forestry Agency, Japan (1976) 'Managing forests for water supplies and resource conservation'. *Hydrological Techniques for Upstream Conservation*. F.A.O. Conservation Guide No. 2. Rome: Food and Agriculture Organisation of the United Nations.

Galbraith, A.F. (1975) 'Method for predicting increases in water yield related to timber harvesting and site conditions'. *Watershed Management Symposium, Logan, Utah*. New York: American Society of Civil Engineers, Irrigation and Drainage Division.

Gash, J.II.C. (1978) 'Comment on the paper by Thom, A.S. and Oliver, H.R. "On Penman's equation for estimating regional evaporation" '. *Quarterly Journal of the Royal Meteorological Society* **104**, 532-33.

Gash, J.H.C. (1979) 'An analytical model of rainfall interception by forests'. *Quarterly Journal of the Royal Meteorological Society* **105**, 43-55.

Glover, J. and McCulloch, J.S.G. (1958) 'The empirical relationship between solar radiation and hours of sunshine'. *Quarterly Journal of the Royal Meteorological Society* **84**, 172-75.

Gray, D.M. (1970) *Handbook on the Principles of Hydrology*. Port Washington, New York: Water Information Center, Inc., National Research Council of Canada.

Gray, D.M. and Male, D.H. (1981) *Handbook of Snow: Principles, Processes, Management and Use*. Toronto: Pergamon.

Grindley, J. (1967) 'The estimation of soil moisture deficits'. *Meteorological Magazine* **96**, 97-108.

Grindley, J. (1969) *The Calculation of Actual Evaporation and Soil Moisture Deficit over Specified Catchment Areas*. Hydrological Memorandum No. 38. Bracknell, U.K.: Meteorological Office.

Grindley, J. (1972) 'Estimation and mapping of evaporation'. *Proceedings of the Reading Symposium on the World Water Balance*. (July 1970), 200-13. Published jointly by International Association of Scientific Hydrology (Belgium), World Meteorological Organisation (Geneva) and Unesco (Paris).

Hansen, V.E., Israelsen, O.W. and Stringham, G.E. (1980) *Irrigation Principles and Practices*. 4th edn. New York: Wiley.

Heerdegen, R.G. and Reich, B.M. (1974) 'Unit hydrographs for catchments of different sizes and dissimilar regions'. *Journal of Hydrology* **22**, 143-53.

Helvey, J.D. and Patric, J.H. (1965) 'Canopy and litter interception of rainfall by hardwoods of the eastern United States'. *Water Resources Research* **1**, 193-206.

Hendrick, R.L., Filgate, B.D. and Adams, W.M. (1971) 'Application of environmental analysis to watershed snowmelt'. *Journal of Applied Meteorology* **10**, 418-29.

Hibbert, A.R. (1967) 'Forest treatment effects on water yield'. In Sopper, W.E. and Lull, H.W. *eds. Forest Hydrology*. Oxford: Pergamon, 524-44.

Hibbert, A.R. (1981) 'Opportunities to increase water yield in the south-west by vegetation management'. *Proceedings of a Symposium on Interior West Watershed Management* (April 1980). Spokane, Pullman, Washington: Washington State University.

Hicks, C.S. (1975) *Man and Natural Resources: An Agricultural Perspective.* London: Croom Helm.

Hillel, D. (1971) *Soil and Water.* New York: Academic Press.

Holton, H.N., England, C.B. and Whelan, D.E. (1967) 'Hydrologic characteristics of soil types'. *Proceedings of the American Society of Civil Engineers* **93** (IR3), 33-41.

Holtan, H.N. and Lopez, N.C. (1971) *USDAHL-70 Model of Watershed Hydrology.* Technical Bulletin No. 1435. Washington, D.C.: Agricultural Research Service, U.S. Department of Agriculture.

Holtan, H.N., Stiltner, G.J., Henson, W.H. and Lopez, N.C. (1975) *USDAHL-74 Revised Model of Watershed Hydrology: A United States Contribution to the International Hydrological Decade.* Technical Bulletin No. 1518. Washington, D.C.: Agricultural Research Service, U.S. Department of Agriculture.

Hoover, M.D. (1969) 'Vegetation management for water yield'. *Symposium on Water Balance in North America, Banff, Alberta, Canada.* Urbana, Illinois: American Water Resources Association, Proceedings Series 7, 191-95.

Institution of Water Engineers (1969) *Manual of British Water Engineering Practice.* 4th edn. Cambridge: Heffer.

Jeffrey, W.W. (1970) 'Hydrology of land use'. In Gray, D.M. *ed. Handbook on the Principles of Hydrology.* Port Washington, New York: Water Information Center, National Research Council of Canada.

Kuz'min, P.P. (1960) *Formirovanie Snezhnogo Pokrova imetody opredeleniya snegozapasov* [Snow cover and Snow Reserves]. Gidrometeoroligicheskoe, Izdatelitvo, Leningrad. English Translation by Israel Program for Scientific Translations, Jerusalem, Translation No. 828.

Kuz'min, P.P. (1961) *Protsess Tayaniya Snezhnogo Pokrova* [Melting of Snow Cover]. English translation by Israel Program for Scientific Translations, Jerusalem, Translation No. 71.

Lee, R. (1980) *Forest Hydrology.* New York: Columbia University Press.

Leopold, L.B. (1968) *Hydrology for Urban Land Planning – A Guidebook on the Hydrologic Effects of Urban Land Use.* Geological Survey Circular 554. Washington, D.C.: U.S. Department of the Interior.

List, R.J. (1968) *Smithsonian Meteorological Tables.* 6th revised edn. published 1949, fourth reprint. Washington, D.C.: Smithsonian Institution.

Lvovitch, M.I. (1972) 'World water balance (general report)'. *Proceedings of the Reading Symposium on the World Water Balance.* (July 1970) 401-15. Published jointly by International Association of Scientific Hydrology (Belgium), World Meteorological Organisation (Geneva) and Unesco (Paris).

McGowan, M., Williams, J.B. and Monteith, J.L. (1980) 'The water balance of an agricultural catchment. III. The water balance'. *Journal of Soil Science* **31**, 245-62.

Miers, R.H. (1974) 'The civil engineer and field drainage'. *Journal of the*

Institution of Water Engineers **28**, 211-27.

Miller, D.H. (1959) 'Transmission of insolation through pine forest canopy as it effects the melting of snow'. *Mitteilungen der Schweizerischen Anstalt für das forstliche Versuchswesen* **35**, 37-39.

Ministry of Agriculture, Fisheries and Food (1967) *Potential Transpiration*. Technical Bulletin No. 16. London: Her Majesty's Stationery Office.

Ministry of Agriculture, Fisheries and Food (1976) *The Agriculture Climate of England and Wales*. Technical Bulletin No. 35. London: Her Majesty's Stationery Office.

Ministry of Agriculture, Fisheries and Food (1982) *Irrigation*. Technical Bulletin No. 138. London: Her Majesty's Stationery Office.

Molchanov, A.A. (1963) *The Hydrological Role of Forests*. Akademiya Nauk SSSR. Institut Lesa. Israel Program for Scientific Translations, Jerusalem, Catalogue No. 870.

Molz, F.J. and Browning, V.D. (1977) 'Effect of vegetation on landfill stabilization'. *Groundwater* **15**, 409-15.

Monteith, J L. (1965) 'Evaporation and environment'. *Symposia of the Society of Experimental Biology* **19**, 205-34.

Monteith, J.L. (1973) *Principles of Environmental Physics*. London: Arnold.

Musgrave, G.W. and Holtan, H.N. (1964) 'Infiltration'. In Chow, V.T. ed. *Handbook of Applied Hydrology*. New York: McGraw-Hill, Section 12.

Natural Environment Research Council (1975) *Flood Studies Report: Volume 1 Hydrological Studies*. London: Natural Environmental Research Council.

Ogrosky, H.O. and Mockus, V. (1964) 'Hydrology of agricultural lands'. In Chow, V.T. ed. *Handbook of Applied Hydrology*. New York: McGraw-Hill, Section 21.

Oke, T.R. (1978) *Boundary Layer Climates*. London: Methuen.

Penman, H.L. (1948) 'Natural evaporation from open water, bare soil and grass', *Proceedings of the Royal Society of London*. Series A, **193**, 120-45.

Penman, H.L. (1949) 'The dependence of transpiration on weather and soil conditions'. *Journal of the Soil Science* **1**, 74-89.

Penman, H.L. (1951) 'The water balance of catchment areas'. *Association of International Hydrological Science, Brussels Assembly* **3**, 434-42.

Penman, H.L. (1967) 'Comment on the paper by R. Zahner "Refinement of empirical functions for realistic soil-moisture regimes under forest cover" '. In Sopper, W.E. and Lull, H.W. eds. *Forest Hydrology*. Oxford: Pergamon.

Philip, J.R. (1969) 'Theory of infiltration'. *Advances in Hydroscience* **5**, 215-96.

Price, A.G. and Dunne, T. (1976) 'Energy balance computations of snowmelt in a subarctic area'. *Water Resources Research* **12**, 686-94.

Rakhmanov, V.V. (1962) *Role of Forests in Water Conservation*. Goslesbumizdat, Moskva. Translated by Israel Program for Scientific Translations, Jerusalem, Translation No. 1688.

Rantz, S.E. (1971) *Suggested Criteria for Hydrological Design of Storm Drainage Facilities in San Francisco Bay Region, California*. Menlo Park, California: U.S. Geological Survey Open File Report.

Rijtema, P.E. (1965) *An Analysis of Actual Evapotranspiration*. Agricultural Report No. 659. Netherlands: Wageningen.

Russell, G. (1980) 'Crop evaporation, surface resistance and soil water status'. *Agricultural Meteorology* **21**, 213-26.

Rutter, A.J. (1975) 'The hydrological cycle in vegetation'. In Monteith, J.L. ed. *Vegetation and the Atmosphere*. Volume 1. London: Academic Press, 111-54.

Rutter, A.J., Kershaw, K.A., Robins, P.C. and Morton, A.J. (1972) 'A predictive model of rainfall interception in forests, 1. Derivation of the model from observations in a plantation of Corsican pine'. *Agricultural Meteorology* **9**, 367-84.

Salter, P.J. and Williams, J.B. (1965) 'Influence of texture on moisture characteristics of soils. 2. Available-water capacity and moisture release characteristics'. *Journal of Soil Science* **16**, 310-17.

Schwab, G.O., Frevert, R.K., Edminster, T.W. and Barnes, K.K. (1966) *Soil and Water Conservation Engineering*. New York: Wiley.

Shachori, A.Y. and Michaeli, A. (1965) 'Water yields of forest, maquis and grass covers in semi-arid regions: a literature review'. *Arid Zone Research* **25**, 467-77.

Shuttleworth, W.J. and Calder, I.R. (1979) 'Has the Priestley-Taylor equation any relevance to forest evaporation'. *Journal of Applied Meteorology* **18**, 629-646.

Steppuhn, H. (1981) 'Snow and agriculture'. In Gray, D.M. and Male, D.H. *eds*. *Handbook of Snow: Principles, Processes, Management and Use*. Toronto: Pergamon.

Storey, H.C., Hobba, R.L. and Rosa, J.M. (1964) 'Hydrology of forest lands and rangelands'. In Chow, V.T. ed. *Handbook of Applied Hydrology*. New York: McGraw-Hill, Section 22.

Storey, H.C. and Reigner, I.C. (1970) 'Vegetation management to increase water yield'. *Symposium on the Interdisciplinary Aspects of Watershed Management, Montana State University, Bozeman, Arizona*. New York: American Society of Civil Engineers.

Strahler, A.N. (1975) *Physical Geography*. 4th edn. New York: Wiley.

Szeicz, G., Endrödi, G. and Tajchman, S. (1969) 'Aerodynamic and surface factors in evaporation'. *Water Resources Research* **5**, 380-94.

Thom, A.S. and Oliver, H.R. (1977) 'On Penman's equation for estimating regional evaporation'. *Quarterly Journal of the Royal Meteorological Society* **103**, 345-57.

U.S. Army Corps of Engineers (1956) *Snow Hydrology: Summary Report of the Snow Investigations*. Portland, Oregon: U.S. Army Corps of Engineers, North Pacific Division.

U.S. Army Corps of Engineers (1960) 'Runoff from snowmelt'. *Engineering Manual*. No. 1110-2-1406.

U.S. Geological Survey (1967) *Roughness Characteristics of Natural Channels*. Geological Survey Water-Supply Paper 1849. Washington, D.C.: U.S.

Government Printing Office.

U.S. Soil Conservation Service (1970) *Irrigation Water Requirements*. Technical Release No. 21. Washington, D.C.: U.S. Department of Agriculture.

U.S. Soil Conservation Service (1972) 'Hydrology'. *National Engineering Handbook*. Washington, D.C.: U.S. Department of Agriculture, Section 4.

Ward, R.C. (1975) *Principles of Hydrology*. 2nd edn. London: McGraw-Hill.

Water Space Amenity Commission (1980) *Conservation and Land Drainage Guidelines*. London: Water Space Amenity Commission.

Wiesner, C.J. (1970) *Climate, Irrigation and Agriculture*. Sydney: Angus and Robertson.

Zahner, R. (1967) 'Refinement in empirical functions for realistic soil-moisture regimes under forest cover'. In Sopper, W.E. and Lull, H.W. *eds. Forest Hydrology*. Oxford: Pergamon.

Zinke, P.J. (1967) 'Forest interception studies in the United States'. In Sopper, W.E. and Lull, H.W. *eds. Forest Hydrology*. Oxford: Pergamon.

3 Soil stabilisation

3.1 Introduction

Land, the yielder of food and mineral wealth, the harbourer of water, is a major natural resource and a precious asset. As the supporter of life and life-style it is a resource to be nurtured and cherished. Yet it is not. It is a resource used for exploitation — the gleanings providing construction materials, fuel, metalliferous and non-metalliferous ores. More and more, the natural landscape is disfigured, generating gaping holes and spoil tips as monuments to the modern world.

Productive soil formed over aeons is swept, washed, and lost within a few decades. Deserts covering almost a third of the land surface continue to creep forward at an alarming rate — threatening, encroaching and engulfing potentially productive land; reflecting human misuse of a difficult and sensitive environment.

With increasing populations greater portions of land are converted into urban and industrial environments, interconnected by a web of communication routes.

Each land transformation sees the surface stripped of vegetation — exposed and vulnerable to the ravages of wind and water erosion.

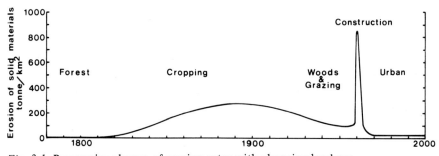

Fig. 3.1 Progressive change of erosion rates with changing land use

The effect of the progressive alteration of the land surface on erosion rates has been documented by Wolman (1967) for a small area near Washington, D.C. A very slow erosion rate of 0.002 m per 1000 y under original forest increased slowly to about 0.1 m per 1000 y as extensive farming with cultivation began to occur in the middle to late nineteenth century (see fig. 3.1). As some of the area returned to forest and grazing land, this rate was reduced by half, only to increase sharply to 10 m per 1000 y as constructional activities and urbanisation began to occur. With complete urbanisation rates dropped again to about 0.01 m or less per 1000 y.

From this brief account we perceive the important influence of vegetation in erosion control, and identify construction activities as a major potential source of sediment. Indeed the equivalent of many decades of natural or even agricultural erosion may take place during a single year from areas for construction. It is not uncommon for such land to remain bare for as long as six months during construction of a single dwelling, or to be exposed for more than one year on larger developments.

Imposition of large quantities of sediment in streams previously carrying relatively small quantities, produce the formation of channel bars; the erosion of channel banks; obstruction of flow; the blanketing of bottom-dwelling flora and fauna causing change; changes in fish populations; the depletion of storage reservoirs, and so on. To avoid such problems, and to control land degradation there is a clear incentive to restore vegetation.

Nature has not only demonstrated how to erode, but also how to protect and there is probably no protective measure initiated by man which has not been invented by nature (Bruun, 1972). By diagnosing situations apparently possessing long-term stability, the engineer can exploit the potential of 'natural' solutions in resolving some of his problems – particularly from the economic point of view. For example, in an emergency spillway, the frequency of occurrence of a flow can be so low that the cost of protective linings, such as blocking work or asphalt, may be too high in relation to the benefits to justify its use. Although there can be more risk with a design relying on vegetation as a lining medium, the low probability of occurrence of the high flow renders this acceptable, provided the designer has a reasonable appreciation of its properties. Second, there are many situations in which a 'natural' solution is more effective; this is particularly the case where vegetation is used to stabilise gully erosion and sand-dunes. Third, by making use of 'natural' solutions using vegetation, many adverse environmental effects can be minimised and aesthetic values preserved.

Successful exploitation of plants depends on good planning and must be based on an accurate diagnosis of problems, together with a sound knowledge of plants and their role as a stabilising medium. Restoration strategy is considered in section 3.5 and represents a focus within this chapter. It is preceded by an appraisal of some technical aspects of plants as a stabilising medium − showing how various features can be incorporated into analytical models for assessing erosion rates and surface stability. The approach is valuable because it permits a rational assessment of erosion control options and enhances the development of a sound strategy for promoting land stability. In many situations it is vital to maintain a holistic approach. For example, the restoration of gullies focuses not only on the gully, but also on the hydrology of the surrounding area. The same is true of stream-bank protection − wherein one must focus on river stability rather than on corrective measures for an isolated eroding bank.

Because of the range of problems and the diversity of environments inhabited by plants, it is only feasible to outline general principles and to consider a few case situations. In this respect, later parts of the chapter focus on biotechnical methods and earthworks, and on problems involving water as an erosive agent. Other important areas such as planting on industrial tailings or controlling desertification are covered in specialised texts such as Bradshaw and Chadwick (1980) and United Nations (1977) respectively, or in more general texts such as Schiechtl (1980) and Clouston (1977).

3.2 Plant factors in surface stability

Of the many features of plants tending to promote stability, wind attenuation and soil reinforcement are especially important. Plants and wind erosion are considered further in section 3.4.2 and chapter 5. On slopes, root reinforcement is one of many factors controlling surface stability and is examined in section 3.3.2.

3.2.1 Wind speed attenuation
Besides shielding the soil surface from splash erosion (caused by impacting raindrops), aerial parts of the plant fulfil a valuable function in slowing down the wind speed near the ground; this arises because foliar elements protruding into the airflow exert drag, thereby causing the withdrawal of wind momentum and reduction in

wind speed. This phenomenon diminishes the opportunity of surface soil movement and enhances deposition in particle-laden flows. Each of these aspects is critical in the control of wind erosion.

Wind speed in the atmosphere near the ground (but above the foliage height) varies logarithmically with height as described in section 5.2.1. Within a dense plant canopy the variation is often observed to follow an exponential relationship (Bache and Unsworth, 1977) of the form:

$$u(z) = u(h) \exp \left[-\alpha(1 - \frac{z}{h}) \right] \qquad\qquad [3.1]$$

in which $u(z)$ is the wind speed at height z above the ground, h the average height of the plants and α a coefficient dependent on the amount of foliage; α is often in the range 1.5-5.0, increasing with the foliage density. As the vegetation becomes more sparse, the profile shape reverts to the logarithmic form according to the scheme described in Seginer (1974). Salient points to observe are that within dense foliage canopies (typical of developed agricultural crops or a grass sward), windspeeds are very low (~ 0.2 m s^{-1}) and more important, the drag experienced at the surface is negligibly small – thus preventing surface scour.

3.2.2 Root reinforcement
The most obvious way in which plants stabilise soils is by root reinforcement, the root tending to bind the soil and to increase its shear strength.

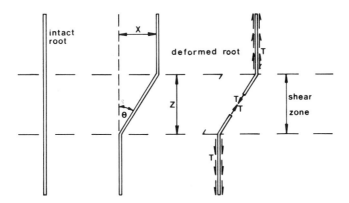

Fig. 3.2 Model of flexible, elastic root extending vertically across a horizontal shear zone

A simple theoretical model of fibre-reinforced soil was developed by Wu *et al.* (1979): the model envisages an elastic root or fibre embedded in the soil and initially oriented normal to the shearing surface (see fig. 3.2). Deformation in the soil is resisted by the tangential forces which develop along the fibre, and ultimately by the tensile strength of the fibre. The angle of internal friction (φ) within the soil is assumed to be unaffected by the reinforcement. From this Wu *et al.* (1979) showed, to a first approximation, that:

$$\Delta S_R \doteq 1.15\, t_R \qquad\qquad\qquad [3.2]$$

where ΔS_R is the shear strength increase from root or fibre-reinforcement and t_R is the average tensile strength of the fibres *per unit area of soil*. The average tensile strength per unit area of soil can be estimated by multiplying the tensile strength of the fibres T_R by the fraction of the shear cross-section filled by fibres:

$$t_R = T_R\, \frac{A_R}{A} \qquad\qquad\qquad [3.3]$$

where A_R is the total fibre cross-sectional area and A the total shear cross-sectional area. If there is a distribution of root sizes, then t_R is determined by a summation of equation 3.3 over all root sizes and their respective strengths. Sometimes it is convenient to represent the root area ratio in terms of the root concentration (R kg m^{-3}) using the substitution $A_R/A = R/\gamma_R$ where γ_R is the actual density (kg m^{-3}) of the root material.

The simple model permits an estimate of the rooting contribution to shear strength based solely on the properties of the roots. The validity of the analysis is supported by results of direct shear tests carried out both in the laboratory and in the field (Endo and Tsuruta, 1968; Waldron, 1977).

The total shear strength S of the soil-root system can be found by adding the shear strength increase, ΔS_R, to the usual Coulomb expression for shear strength:

$$S = (S_s + \Delta S_R) + \sigma \tan \varphi$$

where S_s is the shear strength of the root-free soil and σ the effective normal stress.

According to the fibre-reinforcement model described above, the added strength to the soil-root system is favoured by high root concentrations and roots with a high tensile strength.

The tensile strength of roots has been investigated in a number of studies reported in Gray and Leiser (1982), and Schiechtl (1980). For fresh tree roots the tensile strength appears to depend on root diameter; on slopes, roots growing uphill can be stronger than comparable roots growing downhill. Some woody plants show a seasonal dependence, having the highest tensile strength in late autumn and the lowest in the summer — the time of strongest growth. Burroughs and Thomas (1977) showed that root numbers and tensile strength both decline with age after felling; the consequent reduction in shear strength is an important contributory agent to landslips in logged areas. Tensile strength varies markedly with species. For example, a number of alders and poplars have a mean tensile strength $\gtrsim 3.43 \times 10^4$ kN m^{-2} contrasting with $T_R \sim 1.47 \times 10^4$ kN m^{-2} for willows. For corresponding root diameters, the strength of trees is roughly 1.5-3 times greater than that of the roots of grasses.

To fully exploit the fibre-reinforcement model we also require estimates of the root area ratio. Unfortunately there have been few studies in which the root structure distribution and the strength have been measured simultaneously. Data from Burroughs and Thomas (1977) indicates an area ratio in the range 0.05-0.15% as representative of the rooting systems of Douglas fir. Coupled with tensile strength measurements, this yields a corresponding shear stress increase in the range 3.4 kN m^{-2} to 17.2 kN m^{-2}. From a series of order of magnitude calculations, Gray (1978) argued that increases of this order can be critical in maintaining the stability of slopes with decomposed granitic soils which otherwise possess little cohesion.

3.3 Role of woody vegetation in stabilising slopes

Downslope soil movement by landslides and earth flows is a significant erosional process in steep-sloped catchments in many regions. Besides the mass-wasting of valuable soil, they cause increased sediment load in rivers. For example, in the Eel River watershed of northern California an estimated 25% of sediment in streams is from landslides (Waldron, 1977). Many surveys, e.g. Forestry Agency, Japan (1976) have shown that they are more prevalent in non-forested areas and are associated with logging operations, fire or brush removal — each contributing to the removal of cover. From such evidence it is generally accepted that a forest cover is effective in reducing the incidence of landslides.

Logging or destruction of vegetation cover is most likely to trigger shallow slides, such as debris avalanches; these are the ones over which the forest manager can exert most control. Sound forest management to minimise landslide risk must be based on an understanding of the underlying causes and processes, though one may never eliminate the random factor. Here we shall consider the factors associated with vegetative cover and its removal, on a theoretical basis; related managerial aspects are reviewed in Rice (1977).

3.3.1 General considerations

The stability of a slope depends on a delicate balance between forces. In general, slopes fail when the shear stress on any potential failure surface exceeds the shear strength. It is customary to express this balance of forces in terms of a factor of safety against sliding, and is commonly defined as follows:

$$F = \frac{\text{shear strength along the surface}}{\text{shear forces promoting sliding on a critical surface}} \qquad [3.5]$$

A safety factor of 1 would indicate imminent or incipient failure. It is essential to note that this factor is calculated mainly to assess the danger of sudden or catastrophic failure. Slow yielding or creep may take place even at factors of safety greater than 1.

Possible ways in which vegetation might affect the balance of forces are: mechanical reinforcement from the root system; slope surcharge from the weight of the trees; wind leverage and root wedging; modification of the soil moisture distribution and pore water pressures; and lateral restraint by buttressing and soil arching (see Gray, 1970; Gray and Leiser, 1982).

With the exception of wind leverage and root wedging, each of these factors generally enhances stability. At first sight, the surcharge would appear to increase shear stress, but this effect is largely negated by the accompanying increase in shear strength. Interception and transpiration as modes of water loss each contribute to the reduction of soil moisture. Moisture depletion not only reduces the unit weight of soil, but also enhances cohesion due to the surface tension forces in partially saturated soils. Buttressing and arching refer to the lateral restraint on soil movement from the trunks and roots. Arching in slopes occurs when soil attempts to move through and around a row of piles (or trees) firmly embedded or anchored in an unyielding layer.

Wind leverage or wind throw can be a serious problem caused by the overturning moment of wind on trees, or arising from excessive vibrations which cause loosening of the roots. Root wedging is an alleged tendency of roots to penetrate soil, thereby loosening it or opening cracks.

3.3.2 *Stability analysis*

To gauge the interaction of all these factors on slope stability, we shall consider the regime illustrated in fig. 3.3 depicting a shallow soil layer supporting trees on an infinite slope. The potential failure surface is assumed to be at the interface of the soil and the bedrock surface.

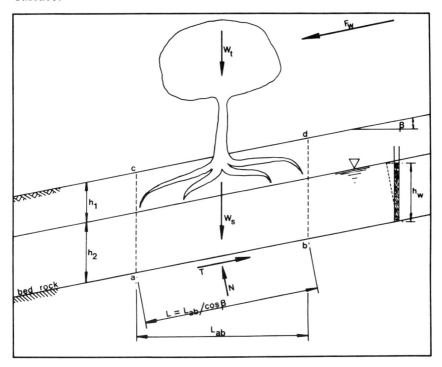

Fig. 3.3 Forces on a sliding soil mass

Under conditions of steady seepage the factor of safety may be written:

$$FS = \frac{S_s + \Delta S}{T} \qquad\qquad [3.6]$$

where S_s is the shearing resistance of soil over the area $L \times 1$ (unit width), ΔS, the increased shearing resistance due to root reinforcement and T, the total shear forces on the failure plane that promotes downslope movement of the soil mass. From fig. 3.3 the component forces downslope are:

$$T = (W_s + W_t) \sin \beta + F_w \qquad [3.7]$$

in which W_s is the bulk unit weight of soil, W_t the added surcharge due to the weight of the trees, F_w the shear force induced by the wind loading on the trees and β the slope angle. Using equation [3.4] to define the shear strength term $S_s + \Delta S$, the factor of safety was expressed by Wu *et al.* (1979) in the form:

$$FS = \frac{(c' + \Delta s)L + (W_s' + W_t) \cos \beta \tan \varphi'}{(W_s + W_t) \sin \beta + F_w} \qquad [3.8]$$

where W_s' signifies the apparent weight of the soil taking account of buoyancy, c' the effective cohesion of root-free soil, Δs the shear strength increment per unit area of soil (*note* $\Delta S = \Delta sL$) and φ' is the effective angle of internal friction of the soil. Expressing the forces in terms of unit weights and taking the width of the column as unity, gives:

$$\left.\begin{array}{l} W_s' = \gamma_1 h_1 + (\gamma_2 - \gamma_w)h_2 \\ W_s = \gamma_1 h_1 + \gamma_2 h_2 \\ W_t = P_t/\cos \beta \\ F_w = \tau_w/\cos \beta \\ L = L_{AB}/\cos \beta \quad (L_{AB} = 1 \text{ as unit length}) \end{array}\right\} \qquad [3.9]$$

where P_t is the weight of trees per unit area of slope, τ_w the shear stress from the wind, γ_1 and γ_2 the unit weight of soil above and below the water table respectively, γ_w the unit weight of water (9.81 kN m^{-2}) together with h_1 and h_2 representing the level of the water table below ground level and above the failure plane respectively.

3.3.3 Effect of cutting trees

SOIL MOISTURE REGIME

Many studies have shown that soil on cut-over lands often possesses a higher moisture content than that on forested lands and that this is largely attributable to the relative evapotranspiration rates (fig. 3.4).

Generally, the drier the climate, the more pronounced is the difference. But if both a forested area and a clear-cut area are near field capacity during most of the year, the effect of timber harvesting may be negligible.

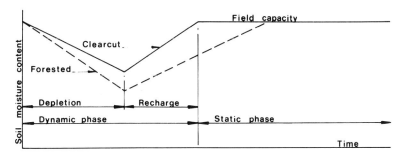

Fig. 3.4 Idealised representation of soil moisture cycle in forested and clear cut areas

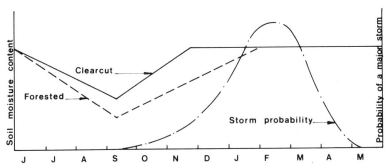

Fig. 3.5 Idealised soil moisture deficit which does not coincide with period of landslide-producing storms

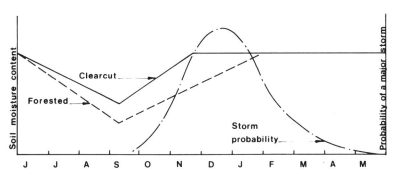

Fig. 3.6 Idealised soil moisture deficit which coincides with landslide-producing storms

As far as landslides are concerned, it is the amount and the duration of the deficit which are important (Rice, 1977). If the period of deficit does not extend into the portion of the rainy season when landslides are most likely to occur (fig. 3.5), then the difference in soil moisture may be of little practical consequence. If the deficit period coincides with the period of maximum landslide stress (fig. 3.6), then the soil moisture depletion effect of the timber harvest may be crucial and lead to greater numbers of landslides in a clear-cut area. As regrowth occurs, the soil moisture depletion effect diminishes; its duration will depend upon the vigour and density of the vegetation that follows the timber harvest.

Table 3.1   Effect of vegetation on stability factor for an infinite slope with steady seepage. Calculations are based on equation [3.8] in conjunction with fig. 3.3 (Charutamra, 1981)

Fixed data
$c' = 5.3$ kN m^{-2} } consistent with
$\varphi' = 34.7$ kN m^{-2} } Wu *et al.* (1979)
$\gamma_1 = 18.0$ kN m^{-3}
$\gamma_2 = 20.0$ kN m^{-3}
$\gamma_w = 9.8$ kN m^{-3} } assumed
$\beta = 35°$
$h_1 + h_2 = 0.9$ m
$L = (1/\cos 35) = 1.22$ m

	h_1 (m)	h_2 (m)	ΔS (kN m^{-2})	P_t (kN m^{-2})	τ_w (kN m^{-2})	FS
Case 1 uncovered slope	0.4 (assumed)	0.5 (assumed)	0.0 (no plant	0.0 on slope)	0.0	1.36
Case 2 plant growth	0.7 (assumed)	0.2 (assumed)	6.0	3.8 (Wu *et al.* 1979)	0.1	1.93
Case 3 dense growth	0.9 (maximum)	0.0	6.0	3.8	0.1	2.18
Case 4 cut-over immediate	0.9	0.0	6.0 (fresh roots)	0.0 (plants removed)	0.0	2.90
Case 5 after cut-over	0.4 (case 1)	0.5 (case 1)	3.0 (50%)	0.0	0.0	1.75
Case 6 long after cut-over	0.0	0.9 (maximum)	0.0	0.0	0.0	1.13

ROOT DECAY

Under most circumstances much of the increase in landslides after timber harvest can be attributed to the decrease in the strength of the slope from root decay (Rice, 1977). For example, Burroughs and Thomas (1977) showed that numbers of roots per unit area of soil and their individual tensile strength declined with time in both Coast and Rocky Mountain varieties of Douglas fir. By combining these features, it was calculated that within 30 months after felling, Coast Douglas fir lost 86% of its total root tensile strength per unit area of soil and Rocky Mountain Douglas fir lost 65% of its strength.

STABILITY CALCULATIONS

To illustrate the interplay of these factors on slope stability, the factor of safety may be calculated using equation [3.8] for a variety of situations prior to and following cutting. In table 3.1, cases 1 to 3 show the factor of safety increasing as the soil moisture content is reduced – perhaps attributable to the degree of plant cover. In case 4 cut-over is assumed and the factor of safety attains a maximum value commensurate with conditions of maximum root reinforcement, soil moisture at its lowest level and the effect of wind leverage removed. Cases 5 and 6 exemplify the gradual reduction of root strength and a corresponding increase in soil moisture. The slope is at its most vulnerable when the factor of safety is least (case 6) coinciding with saturated soil conditions and absence of root reinforcement.

3.4 Erosion analysis

Erosion may be considered in two categories: geological or natural erosion, and accelerated erosion. Geological erosion occurs on undisturbed soils in their natural state and is not considered to be a problem, but an essential part of forming the normal landscape. This type of erosion is responsible for soil formation and through time has produced most natural topographic features. Under conditions of natural erosion, soil formation and erosion are considered to be in a state of balance so that the soil thickness remains fairly constant over time. The natural vegetation cover is of vital importance in the maintenance of this balanced state and anything which disturbs it tends to produce accelerated erosion.

Considerable research into the underlying processes of accelerated erosion has led to the development of analytical models for assessing erosion rates and quantifying management techniques; these are considered below, identifying the contributory role of vegetation.

3.4.1 Water erosion

Water erosion takes different forms: splash erosion caused by the impact of raindrops; rills in which water is concentrated in small rivulets; gully erosion where the eroded channels are larger; and finally stream-bank erosion when rivers or streams are cutting into their banks. On arable land, splash and rill erosion are the most important. On sloping land, the concentration of water flow into rills increases its erosive power and rill erosion may account for the bulk of sediment removed from a hillside. Where there are high sediment loads in streams and rivers which threaten to silt up storage dams, the most important source of this silt is probably gully erosion or stream-bank erosion (Hudson, 1971).

PROCESSES

Soil erosion may be regarded as a two phase process consisting of, first, the detachment of individual particles from the soil mass, then their transport by erosive agents such as running water and wind. When sufficient energy is no longer available, deposition occurs.

Rain-splash is the most important detaching agent due to the magnitude of the incipient kinetic energy. The transference of momentum to the soil surface has two effects. First, it provides a consolidating force leading to surface compaction and second, it mobilises soil particles — which on a sloping surface leads to a net downhill transport. When the rainfall rate exceeds the infiltration capacity of the soil, water accumulates on the soil surface and overland flow may occur. If the flow velocity exceeds some threshold value, further erosion will occur dependent on the grain size and flow velocity. An interaction exists between splash erosion and surface wash, in that surface sealing associated with splash erosion reduces infiltration — thereby increasing the overland flow and exacerbating surface scour.

In view of these basic processes, the principal effects of vegetation in reducing erosion may be listed as:
(a) the interception of the raindrops so that their kinetic energy is dissipated by plants rather than imparted to the soil;
(b) increasing the frictional losses associated with surface flows (see section 2.6) and thereby reducing the erosive potential and the transport capability;
(c) increasing the infiltration capacity of the soil thereby reducing

the probability of overland flow (see section 2.3);

(d) the physical restraint of soil movement.

MODELLING SOIL EROSION

The most rudimentary models are of a 'black box' type — relating sediment loss to either rainfall or runoff. A typical relationship is:

$$Q_s = a \, Q^n \qquad\qquad\qquad [3.10]$$

For example, Fleming (1969) analysed data from 250 drainage basins in the world and yielded the information shown in table 3.2. The data provides a simple ranking of the cover types and shows that (undisturbed) forest offers the best protection against erosion.

Table 3.2   Values of a and n in equation [3.10] for various cover types

Cover type	a	n
Mixed broad-leaf and coniferous	117	1.02
Coniferous and tall grassland	3 523	0.82
Short grassland and scrub	19 260	0.65
Desert and scrub	37 730	0.72

Although many of the contributory processes can be evaluated individually, their synthesis into sophisticated models such as the Stanford Sediment Model (Gregory and Walling, 1973) is cumbersome and usually remote from common engineering practice. Alternatively, advantage can be gained by using macroscopic models, e.g. the Universal Soil Loss Equation, which, despite its over-simplistic and empirical nature, provides some insight into the way different variables affect sediment yield; it also offers scope for the development of an erosion control strategy.

UNIVERSAL SOIL LOSS EQUATION

Studies conducted by the U.S. Soil Conservation Service and others have shown that the erosion caused by rain is essentially a function of the erosivity of the rain and the erodibility of the soil. The erosivity of the rain depends on its incipient energy, whereas the erodibility of soil is determined by a number of contributory factors which can be subdivided into two groups. The first contains those factors which reflect the soil character, i.e. its mechanical, chemical and physical composition. The second group of factors depends on how the soil is managed, embracing both land management and crop management – though it is sometimes difficult to distinguish between these. The kind of land use, e.g. forestry, grazing, arable, etc., comes under the heading of land management. These however, embody factors such as kind of crop, fertiliser treatment, harvesting and so on – which are facets of crop management. Some conservation practices such as contour ploughing or terracing, are bound up with the broad issues of land management and the mechanics of crop management.

The combination of the various components of erosivity and erodibility has led to the development of the Universal Soil Loss Equation (Wischmeier and Smith, 1962) shown below:

Erosion is a function of:

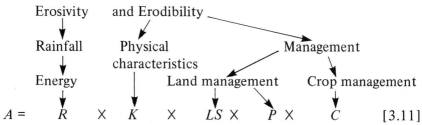

$$A = \quad R \quad \times \quad K \quad \times \quad LS \times \quad P \times \quad C \qquad [3.11]$$

This computes the soil loss A as a product of the rainfall erosivity index R, a soil factor K, slope length L and slope steepness S as a single index LS, an erosion control practice factor P and a cropping management factor C.

In the f.p.s. system of units (which is normally used) A is measured in tons per acre (short tons of 2000 lb) over the period of interest.

The rainfall erosivity index is a measure of the erosive force of specific rain. This is calculated from the total kinetic energy (E) of each rainstorm multiplied by its maximum intensity over 30 minutes (I_{30}), divided by 100. The kinetic energy varies with the rainfall intensity according to the relationship:

$$E = 916 + 331 \log_{10} I \qquad [3.12]$$

where I is the rainfall intensity (in h^{-1}) and E is the kinetic energy expressed in foot-tons per acre per inch of rainfall.

To compute the kinetic energy of a storm, a trace of the rainfall from a recording rain gauge is analysed and the storm divided into increments of uniform intensity. For each period of time, knowing the intensity of the rain, E can be estimated from the above

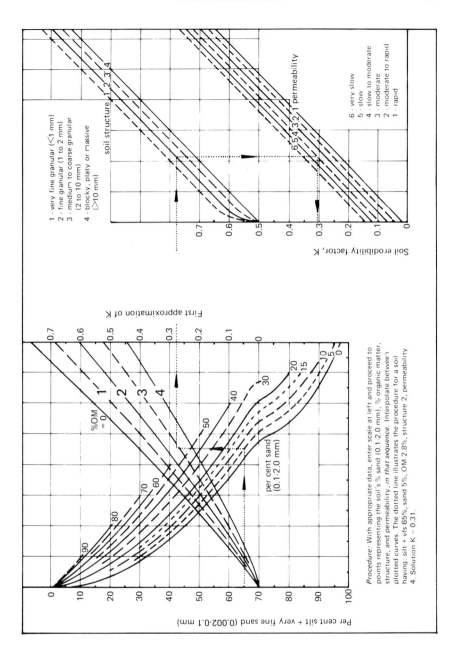

Procedure: With appropriate data, enter scale at left and proceed to points representing the soil's % sand (0.1-2.0 mm), % organic matter, structure, and permeability, in that sequence. Interpolate between plotted curves. The dotted line illustrates the procedure for a soil having: silt + vfs 65%, sand 5%, OM 2.8%, structure 2, permeability 4. Solution K = 0.31.

Fig. 3.7
Soil erodibility
nomograph

equations and this, multiplied by the amount of rain received, gives the kinetic energy for that time period. Increments of energy are then summed to obtain E for the whole storm. The rainfall erosivity index can then be calculated from the sum of factor $EI_{30}/100$ for each storm during the period of interest.

According to Hudson (1971) erosion is caused almost entirely by rain falling at intensities greater than some critical level (25 mm h^{-1}). Hudson proposed an alternative erosivity index based on this assumption; this is considered to be more suitable for tropical and subtropical regions with high rainfall intensities.

The soil erodibility factor represents the soil loss per unit of EI_{30} as measured on a standard bare soil plot, 22 m long and of 5° slope. K may be estimated from a nomograph (fig. 3.7) provided the grain size distribution, organic matter content, structure and permeability are known.

The factor LS can be calculated from nomographs (Wischmeier and Smith, 1978) or estimated from the equation:

$$LS \doteq L^{0.5} \times (0.0076 + 0.0053S + 0.00076S^2) \qquad [3.13]$$

where L is in ft and S is a percentage.

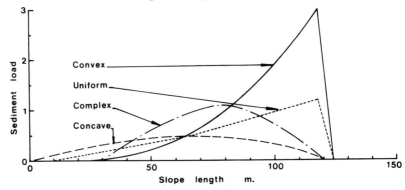

Fig. 3.8 Sediment load or total erosion (tonnes per metre of slope width) along the four original slope shapes of 5% steepness during their first erosion period. The sediment load was increasing and net erosion taking place where the curves have positive slopes. Deposition occurred where curves have negative slopes. The steepness at any point along a curve indicates the depth of erosion or deposition

When a slope is concave, convex or irregular, the average steepness does not accurately predict slope effect. The soil-loss rate near the top of a convex slope (steepening toward the bottom) is greater than on a slope of equal elevation change (Meyer and Kramer, 1969). On concave slopes it is less; these relationships are shown in fig. 3.8 and can be very significant in reducing sediment yields from re-shaped lands. The magnitude of the effect of the curvilinearity can be approximated by dividing the slope into segments.

Table 3.3   *P* factors for contouring, contour strip cropping and terracing

| Land slope | | P values | | |
| | | Contour | Terracing | |
(%)	Contouring	strip cropping	*	**
2.0 to 7	0.50	0.25	0.50	0.10
8.0 to 12	0.60	0.30	0.60	0.12
13.0 to 18	0.80	0.40	0.80	0.16
19.0 to 24	0.90	0.45	0.90	0.18

*For erosion control planning on farmland
**For prediction of contribution to off-field sediment load

The conservation practice factor is obtained from tables of the ratio of soil loss where conservation measures such as contouring, strip cropping and terracing are practised, to that where they are not (see table 3.3). With no conservation practices $P = 1.0$; otherwise representative values are shown in table 3.3.

The cropping management factor describes the total effect of vegetation, residue, soil surface and management, on soil loss. It represents the ratio of soil loss under a given crop to that from bare soil.

Values of C are given in Wischmeier and Smith (1962) for many different crop rotations and stages of growth. The U.S. Soil Conservation Service (1975) also provides C values for other agricultural regions, and for woodland and pasture as shown in tables 3.4 and 3.5. C ranges from near zero for excellent sod to 1.0 for continuous fallow. On construction sites C reflects the influence of various types and rates of mulch, methods of revegetation, chemical soil stabilisers and loose and compacted fills (see Wischmeier and Meyer, 1973; Wischmeier and Smith, 1978).

The Universal Soil Loss Equation (U.S.L.E.) has been derived from extensive data collected within the United States and is widely accepted as a reliable tool for analysing erosion from arable land. Although designed for the special conditions of America it represents a logical structure which can be adapted for other countries and to different land uses. It attempts only to predict splash, sheet and rill erosion with no consideration of gully erosion or stream-bank erosion, etc. It predicts just the amount of soil moved from its original position on a field or hillside and does not estimate the soil erosion, i.e. resulting from the difference between deposition and erosion.

Table 3.4   *C* factors for woodland

Stand condition	Tree canopy* (% of area)	Forest litter** (% of area)	Undergrowth†	C factor
Well stocked	100 − 75	100 − 90	Managed††	0.001
			Unmanaged††	0.003 − 0.011
Medium stocked	70 − 40	85 − 75	Managed	0.002 − 0.004
			Unmanaged	0.01 − 0.04
Poorly stocked	35 − 20	70 − 40	Managed	0.003 − 0.009
			Unmanaged	0.02 − 0.09§

 *When tree canopy is less than 20%, the area will be considered as grassland, or cropland for estimating soil loss. See table 3.5.
 **Forest litter is assumed to be at least 2 inches deep over the percentage ground surface area covered.
 †Undergrowth is defined as shrubs, weeds, grasses, vines, etc. on the surface area not protected by forest litter. Usually found under canopy openings.
 ††Managed − grazing and fires are controlled.
 Unmanaged − stands that are overgrazed or subjected to repeated burning.
 § For unmanaged woodland with litter cover of less than 40%, *C* values should be taken from table 3.5.

Because of its simple structure the equation can be re-arranged so that if an acceptable value of *A* is chosen, the slope length required to reduce soil loss to that value can be calculated. Similarly a *C* or *P* value can be chosen and related tables may be scanned to discover the most appropriate conservation practices commensurate with the selected value.

Although the U.S.L.E. was designed for quantifying soil losses from arable land, its use can be viewed in a wider context. For example, if we assume a given nutrient concentration in the surface soil, the U.S.L.E. can be used to estimate the basic soil loss and thereafter, the associated nutrient loss from farm fields under different cropping and management practices (see Holt *et al.* 1979). In consequence one may gain knowledge of the potential quality of surface waters discharged from agricultural land. Wischmeier and Meyer (1973) discuss the use of the U.S.L.E. for assessing erosion control strategies on construction sites, an aspect which is also pursued in Dunne and Leopold (1978).

Table 3.5 *C* values for permanent pasture, rangeland and idle land*

Type and height of raised canopy**	Canopy cover† (%)	Type††	Cover that contacts the surface (Percentage ground cover)					
			0	20	40	60	80	95-100
No appreciable canopy		G	0.45	0.20	0.10	0.042	0.013	0.003
		W	0.45	0.24	0.15	0.090	0.043	0.011
Canopy of tall weeds	25	G	0.36	0.17	0.09	0.038	0.012	0.003
or short brush		W	0.36	0.20	0.13	0.082	0.041	0.011
(0.5 m fall height)	50	G	0.26	0.13	0.07	0.035	0.012	0.003
		W	0.26	0.16	0.11	0.075	0.039	0.011
	75	G	0.17	0.10	0.06	0.031	0.011	0.003
		W	0.17	0.12	0.09	0.067	0.038	0.011
Appreciable brush	25	G	0.40	0.18	0.09	0.040	0.013	0.003
or bushes		W	0.40	0.22	0.14	0.085	0.042	0.011
(2 m fall height)	50	G	0.34	0.16	0.085	0.038	0.012	0.003
		W	0.34	0.19	0.13	0.081	0.041	0.011
	75	G	0.28	0.14	0.08	0.036	0.012	0.003
		W	0.28	0.17	0.12	0.077	0.040	0.011
Trees but no appreciable	25	G	0.42	0.19	0.10	0.041	0.013	0.003
low brush		W	0.42	0.23	0.14	0.087	0.042	0.011
(4 m fall height)	50	G	0.39	0.18	0.09	0.040	0.013	0.003
		W	0.39	0.21	0.14	0.085	0.042	0.011
	75	G	0.36	0.17	0.09	0.039	0.012	0.003
		W	0.36	0.20	0.13	0.083	0.041	0.011

*All values shown assume: (a) random distribution of mulch or vegetation, and (b) mulch of appreciable depth where it exists.

**Average fall height of waterdrops from canopy to soil surface: m = metres.

†Portion of total area surface that would be hidden from view by canopy in a vertical projection (a bird's-eye view).

††G: Cover at surface is grass, grasslike plants, decaying compacted duff, or litter at least 5 cm (2 inches) deep. W: Cover at surface is mostly broad-leaf herbaceous plants (such as weeds) with little lateral-root network near the surface, and/or undecayed residue.

Besides the development of a conservation strategy, it is useful to gain estimates of the net soil erosion. Sediment yield is often determined by applying a delivery ratio to an estimate of the gross erosions. Sediment Delivery Ratio (S.D.R.) is defined as the ratio of the sediment delivery to gross erosion on the catchment. Generally the ratio is less than one due to deposition on land surfaces and in streams. Gross erosion is generally estimated using the U.S.L.E. combined with estimates for other sources such as channels and gulleys, etc. The S.D.R. is usually estimated on the basis of catchment size:

$$\text{S.D.R. } (\%) = 36 \, A^{-0.20} \qquad\qquad [3.14]$$

where A is the drainage area in km^2 (Roehl, 1962), together with adjustments for land use, topography, etc. within the area (see American Society of Civil Engineers, 1975).

Example 3.1 Universal soil loss equation

Estimate the average annual soil loss from disturbed land which is initially bare and subsequently revegetated. The characteristics of the area are as follows:

Area: 24 acres
Average hillslope length: 300 ft
Average hillslope gradient: 10%
Rainfall erosivity: 25 000 foot-tons per acre per year
Soil: 65% silt and very fine sand; 5% sand; 2.8% organic matter; fine granular structure and slow to moderate permeability.
Cover: Bare initially and eventually covered by a combination of tall weeds (0.5 m fall height) with 25% canopy cover and grass covering 80% of the surfaces.
Conservation practices: None
$R = 25\,000/100 = 250$
$K = 0.31$ (from fig. 3.7)
$LS = 2.37$ (from equation [3.13])
$C = 1.0$ (bare), 0.012 (vegetated) (from table 3.5)
$P = 1.0$

From equation [3.11]:
(bare)

$A = 250 \times 0.31 \times 2.37 \times 1.0$
$\quad = 184 \text{ tons acre}^{-1} \text{ y}^{-1}$
i.e. soil lost from bare site $= 184 \times 25$
$= 4600 \text{ tons y}^{-1}$

(vegetated)

$A = 184 \times 0.012$
$\quad = 2.2 \text{ tons acre}^{-1} \text{ y}^{-1}$
i.e. soil lost from vegetated site $= 2.2 \times 25$
$= 55 \text{ tons y}^{-1}$

3.4.2 Wind erosion

The severity of wind erosion depends on equilibrium conditions between soil, vegetation and climate. It is accelerated by processes that cause disintegration in the surface soil structure and the depletion of vegetative cover. Conversely, wind erosion is hindered by stabilisation processes such as soil consolidation and aggregation, and by vegetation and residues on the surface.

The design of suitable methods of controlling wind erosion depends on knowledge of the contributory factors, which are described in an extensive review by Chepil and Woodruff (1963).

PROCESSES

Movement of soil is initiated when the pressure of the wind against the surface of the soil grains overcomes the constraining forces on the soil grains. Three distinct kinds of movement occur: suspension, surface creep and saltation, the latter referring to grains moving along the surface in a series of jumps. Of the three mechanisms, saltation is by far the most important because of its dominance in mass transport and also the fact that neither creep nor suspension occur without saltation (Hudson, 1971). This view was emphasised in Chepil and Woodruff (1963) who stated that wind erosion occurs only when soil grains capable of being moved in saltation are present in the soil. Over 90% of the saltating grains do not rise higher than 0.3 m; thus wind erosion is essentially a surface phenomenon extending to the saltation height.

Chepil and Woodruff (1963) indicated that the most erodible particles of specific gravity 2.65 are about 0.1 mm diameter; in dry conditions these require a wind speed of about 4.5 m s^{-1} at 0.3 m over a smooth surface (corresponding to $z_0 = 2 \times 10^{-6}$m, $u_* = 0.15$ m s^{-1} in equation [5.1]) to initiate movement. However threshold wind speeds may be considerably greater, depending on other factors (Skidmore and Woodruff, 1968).

When the wind speed is greater than the threshold required to initiate soil movement then q, the rate of movement (weight transported per unit width per unit time), is proportional to the wind speed cubed. For wet soil, q varies inversely as the square of the effective soil moisture.

The roughness of the ground surface also has a considerable influence on erosion due to its effect on the wind profile near the ground. Inspection of equation [5.1] shows that as the surface roughness increases the wind speed at a given height is reduced. In addition, a rough surface of clods, ridges and furrows or with vegetation, offers greater potential than a smooth surface for trapping erodible fractions.

On an unprotected field $q = 0$ on the windward edge and increases with distance downwind, until (if the field is large enough) the flow becomes a maximum consistent with the particular wind speed and soil properties.

Of the surface character, the importance of vegetative protection on the land cannot be over-stressed and led Chepil and Woodruff to comment that the greatest frequency and magnitude of wind erosion occurs on soils that have been partly or completely denuded of vegetation or surface cover. Beside soil binding (see section 3.2.2) the dominant facet of vegetation is its ability to slow down the surface wind speed due to the drag exerted by foliar elements. If the foliage density (foliage area per unit volume) is sufficiently great, the surface wind speed is reduced to a negligible level (~ 0.2 m s^{-1} as illustrated in Bache and Unsworth (1977). If the vegetation is less dense, these effects are less pronounced (see section 3.2.1).

A related aspect is the association between vegetative cover and the effective aerodynamic roughness of surface. Data in table 5.1 shows that the roughness length z_0 increases with h, the height of the foliage elements and may be roughly estimated from the approximation $z_0 = h/10$. A greater surface roughness implies that larger amounts of energy are dissipated near the ground and this manifests as a reduction in the surface wind speed; these factors are discussed more fully in section 5.2.1.

MODELLING WIND EROSION

Of the various attempts to predict the degree of erosion, one of the most comprehensive is the Wind Erosion Equation developed by W.S. Chepil on the basis of nearly 30 years of research in the United States. Its origins are described in Chepil and Woodruff (1963) and take the form:

$$E = \mathrm{f}(I, C, K, L, V) \qquad [3.15]$$

where E is the soil loss by wind erosion; I, the erodibility; C, a climatic factor; K, a roughness factor; L, the maximum unsheltered distance across a field in the prevailing wind direction; and V is a measure of the vegetative cover (usually expressed in lb (dry weight)/acre). The mathematical relationship between these components is complex. However graphical and tabular solutions are given in Skidmore and Woodruff (1968) and Woodruff and Siddoway (1965). Besides predicting soil losses, equation [3.15] is extremely useful for assessing the management of wind erosion.

WIND EROSION CONTROL
Erosion can be controlled by a number of different measures including: (a) trapping of moving soil particles; (b) consolidation and aggregation of soil particles; (c) revegetation of the surface; and (d) reducing the wind speed in the atmosphere near the ground.

Soil trapping
The trapping of eroding particles is known as the stilling of erosion and is regarded as an emergency method since its effects are only temporary. Trapping may be accomplished by roughening the surface, by burying the erodible fractions or by placing barriers in the path of the wind. Barrier strips are often used in agricultural regions prone to erosion and consist of small grain crops or grasses planted at various distances apart. They are usually designed to prevent saltating particles from jumping over them; this determines their geometry and they should also possess sufficient storage capacity to hold the soil they trap. From analysis of field and wind tunnel tests Chepil (1960) developed an empirical procedure for determining the maximum width of field strips to control wind erosion and this is also discussed in Schwab *et al.* (1966) and Woodruff and Siddoway (1965).

The latter approach is based on equation [3.15] which can be used to design barrier strip systems where the barriers have adequate capacity and 100% trapping efficiency. For narrow strip barriers, the trapping efficiency falls below 100%, and straightforward design procedures to calculate barrier spacings in terms of selected barrier dimensions are described in Hagen *et al.* (1972). Barrier strips also have the advantage of trapping snow and this can help to conserve soil moisture (see section 5.5).

Revegetation
When revegetation is the objective, consideration should also be given to the land use. However, grasses, once established, are usually the most effective because they provide a dense complete cover which lasts all the year round. This contrasts with agricultural crops which have a more pronounced seasonal dependence. Grass that is easily bent in the wind is generally less effective in controlling erosion than grass which is not, because the latter presents a greater surface roughness. For similar reasons, a standing crop or stubble is more effective than flattened vegetation on a weight-for-weight basis.

Reducing wind speed
The three principal methods of reducing surface wind velocities are: vegetative measures, tillage practices and mechanical methods. In each, the degree of protection is influenced by the height and spacing of the obstructions, the braking effect on the wind and the resistance of soil to movement. Tillage practices include roughening the surface by the creation of ridges and furrows at right angles to the prevailing wind direction, though these are effective only to the extent that the roughness elements are non-erodible. Strip cropping with alternate bands of tall and short crops, or leaving unploughed strips of stubble, will also enhance the surface roughness. However a more permanent and efficient means of reducing wind speed lies in the creation of systems of windbreaks and shelterbelts following the principles described in chapter 5.

3.5 Restoration strategy

3.5.1 *General considerations*
When planning the revegetation of land the first priority is to select the eventual land use as this can affect the restoration strategy. For construction activities such as road-building, the endpoint may be a Hobson's choice. Otherwise the restoration of derelict areas may offer a variety of potential land uses, ranging from agricultural and amenity use to land suitable for housing and industry. In some situations the form of the land may restrict its potential use. For example, steeply sloping land is generally inappropriate for growing rotation crops, but may ·be satisfactory for forestry (fig. 3.9). Usually the final choice is strongly influenced by the needs of the local area. The subsequent steps can follow the scheme shown in fig. 3.10.

In any site operation the ideal is to plan the restoration scheme before the commencement of the quest, perhaps by taking account of the siting and shaping of waste materials as well as the type of material involved. It is also advantageous to plan site operations and revegetation procedures in such a way as to minimise the delay between the start of earth-moving and the subsequent restoration. This brings land back into productive use at the earliest stage, reduces the erosion hazard, and minimises the time during which the area remains unsightly.

From the ecological viewpoint, the immediate aim of a rehabilitation project must be to develop a soil ecosystem which permits healthy plant growth and development. A prerequisite is the re-establishment of the natural life and growth cycle of the soil to allow the development of plant and animal life. Organic matter with satisfactory nutrient content and carbon/nitrogen ratio must be present to encourage the development of microorganisms such as fungi and bacteria, since these decompose the organic matter and release the nutrients. At the same time a population of micro-invertebrates such as millipedes (*Diplopoda*) will develop, breaking up the consuming organic matter, thereby enhancing its decay. With this development, one should witness an improvement in the soil structure and an increase in the nutritional status of the suface layer. Usually the process of upgrading the ecosystem takes several years unless a top soil application is used. This development is facilitated by the establishment of pioneer vegetation and paves the way for a natural succession in areas which are suitable for growth. The quickly establishing plants provide soil cover; with good root development and production of biomass, they ensure the development of a dynamic plant community, and thus rapid site improvement.

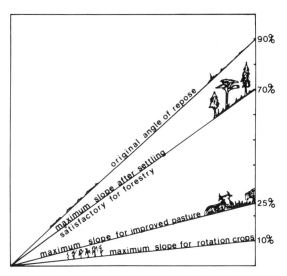

Fig. 3.9 Maximum slopes for various land uses on spoil banks

3.5.2 Fertility problems

Most unproductive soils derived from mining and earth-moving operations are likely to be raw under-developed materials comprising subsoil and rocks. Given time and a suitable climate, weathering will gradually restore the productive potential and a vegetation cover can develop. However it is hardly practical for nature to take its course. Also some spoils pose particular problems which time and nature

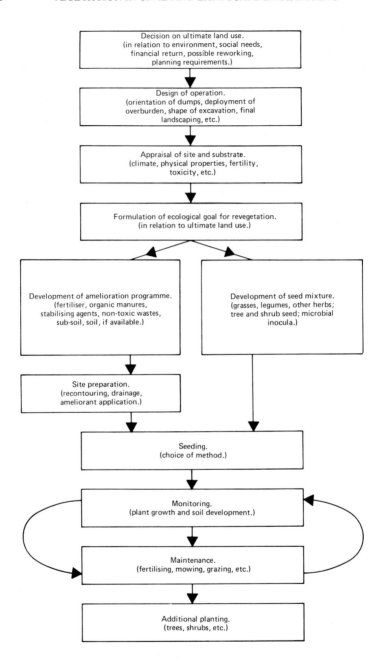

Fig. 3.10 The steps involved in the development of a successful restoration scheme: at each step, careful observations and experiments must be made to ensure that the operations are planned correctly

cannot heal.

In chapter 1 we sketched out the desirable environment of a productive soil and related aspects are discussed in chapter 4. Few derelict land materials have a composition approaching normal soils (see tables 1.4 and 3.6) and frequently suffer from excessive consolidation due to the passage of vehicles (see section 2.3). Invariably they lack organic matter and the associated microbial activity – so vital for the release of nutrients and the formation of a good structure. The surface materials often fail to retain or harbour sufficient moisture to sustain plant growth, but are usually adequate below. Absence of vegetation exposes the substrate to the sun and high surface temperatures may develop. Plants growing sparsely under these conditions are liable to wilt and die. It is rare to find subsurface soils in which there is not a nutrient shortage – tending to be deficient in nitrogen because of the lack of organic matter. Also, if the soil medium lacks an adequate ion exchange capacity, perhaps due to the shortage of humus, there is no way that soluble nutrients such as nitrogen and potassium can be retained. If added, they are likely to be leached out beyond the reach of the plant roots. Problems can also arise from excess acidity, salinity and toxic materials.

Table 3.6   General interpretation of naturally regenerated vegetation on waste substrates

Vegetation character	Interpretations
Degree of vegetational cover:	
absent or sparse	Toxicity, deficiency or physical problems
good cover	Reasonable potential for plant growth
Number of species	A higher number indicates a higher potential for reclamation than a lower number
Species composition	Can indicate substrate conditions, i.e. acidity, salinity, waterlogged, heavy metals
General appearance of plant	Can indicate specific toxicity or deficiency problems
Root appearance:	
long and sparsely branched	Insufficient water supply
shallow-rooted	Toxicity or waterlogging problems

3.5.3 Appraisal of site and substrate

Before any reclamation work can proceed a detailed site investigation is necessary. This takes the form of a progressive series beginning

Table 3.7 Materials can have very different chemical compositions: chemical characteristics of different wastes and degraded areas

	N	P	K	Ca	Mg	pH	Ion exchange capacity Na	Heavy metals	Other toxins
Colliery spoil	○○○	○○○	○	○○○/○	○/	○○○/○ ●●	○○ / ●	○	○
Strip mining	○○	○○○		○○○/○	○/	○○○/○ ●●	○ / ●	○	○
Fly ash	○○○	○	○	○	○	○○○/○ ●●●	○ / ●●●	○/●●	boron
Oil shale	○○○	○	○	○○		○○○/○ ●●	○ / ●●●	○	○
Iron ore mining	○○○	○○○	○	○/	○/	○ ●	○	○	○
Bauxite (strip mining) (red mud)	○○	○○	○	○	○	○	○	○	○
	○○○	○○○	○	○○/○	○/	○○○/ ●●●	○ / ●●●	○	aluminate
Heavy metal wastes	○○○	○○○	○	○○○/	○	●●	○	●/●●●	○
Gold wastes	○○○	○○	○	○	○	○○○	○○	○	○
China clay wastes	○○○	○○○	○	○○	○	○○	○○○	○	○
Acid rocks	○○○	○○○	○	○○	○	○○	○○○	○	○
Calcareous rocks	○○○	○○○	○		○	●	○	○	○
Sand and gravel	○○/○	○/○	○/○	○/○	○/○	○/○	○	○	○
Coastal sands	○○	○	○	○	○	○	○	○	○
Land from sea	○	○	○	○/	○	○/ ●	○ / ●●●	○	○
Urban wastes	○	○	○	○	○	○	○ / ●●	○/●	various
Roadsides	○○	○○	○	○/	○	○/○ ●	○ / ●●	○	○

	deficiency			adequate	excess		
	severe	moderate	slight		slight	moderate	severe
	○○○	○○	○		●	●●	●●●

relative to the establishment of a soil/plant ecosystem appropriate to the material

with a site inspection, and followed by a laboratory analysis of waste materials, pot experiments, field trials and finally site monitoring. Basic site information should also include an appraisal of weather and climatic factors. Pertinent aspects are sketched out below, but are discussed more fully in Bradshaw and Chadwick (1980) and U.S. Environmental Protection Agency (1975). The aim here is to acquaint the reader with the types of tests and procedures which might be carried out; in many instances they are best dealt with by specialists.

SITE INSPECTION

General site characteristics should be noted, paying particular attention to the drainage character, evidence of erosion, variability of substrate and vegetation. Some conditions such as shifting sands, hard rock surfaces or steep and unstable slopes may be immediately apparent, representing conditions in which it is not practical to vegetate unless the basic problems are remedied. The natural regeneration of vegetation can be a very useful guide to the potential of the material for revegetation. Table 3.7 indicates some general features and their interpretation.

SAMPLING

Careful sampling is necessary for planning remedial procedures. If sampling is carried out before commencement of earth-moving it can be useful for planning subsequent management, e.g. burying a particularly adverse material under a more amenable material. Otherwise it is preferable to carry out sampling on a landscaped site.

Sampling for any detailed study should allow for the inherent variability of substrate materials and it is worth stressing that waste materials are nearly always more variable than soil. Two alternative approaches for sampling in a particular area are in common use. These are random sampling and sampling at regular spacings along a transect. The particular system used affects the statistical interpretation of data, but clearly the accuracy of the assessment will improve by increasing the number of samples.

As a rough guide to random sampling before the initiation of earthworks, Gemmell (1977) suggests the numbers of samples taken for analysis should be as follows:

Area (ha)	Number of samples
0 – 4	6
4 – 12	12
12 – 24	15
24 – 40	20

Fewer samples are required when earthworks are complete and as part of the monitoring programme after planting. Each sample should consist of ten or more subsamples taken at random from within a 25 m radius from the sampling point. Once the sampling depth, e.g. 0.15 m has been selected, it is important that the samples consist of equal amounts of material from the surface and at all levels to the chosen depth.

The depth of sampling should be related to the depth of rooting of plants it is hoped eventually to establish. It should also be recognised that the top layers are more subject to change than lower ones.

PHYSICAL ANALYSIS

The most important aspect of a physical analysis is to gain knowledge of the parameters affecting the water status of the surface medium and the following is a rough guide:

water content
water potential (by tensiometer)
infiltration
bulk density
porosity
particle size
surface temperature.

This combination gives insight into mass of water present, the water-holding capacity, the water-release character and the drainage capacity; as well as an indication of the likely consequences of temperature on chemical reactions and microbial activity.

CHEMICAL ANALYSIS

Chemical analysis is essential because nutrient and toxicity problems are universal. It is not possible to provide a universal checklist of the analyses which should be carried out, but those shown in table 3.8 are nearly always required. It is not feasible to measure nutrient availability by chemical tests and the nearest that can be achieved is the measure of extractable ions. The extraction procedure should aim to give an indication of the concentration of ions available to the plant, together with the capacity of the material to keep supplying ions to the solution once some have been removed by the plant roots. Appropriate analytical techniques are described in many texts, e.g. Allen *et al.* (1974). Measurement and estimation of the major macronutrients N, P, K are especially important and it is vital to have a firm grasp of the remaining group, together with micronutrients and any other constituent suspected of causing acidity, toxicity or deficiency problems. Sometimes trace element deficiencies assume greater significance when the land is designated for field crops or grazing purposes.

Table 3.8 Guide to chemical analysis

Parameter	Comment	Reference section
pH	Affects chemical and biological mechanisms, nutrient availability and toxicity	1.5
Exchange acidity (E.A.)	Used to calculate lime requirements	1.5.4
Pyrite and carbonate content	In certain derelict materials indicates long-term trends in acidity	
Cation exchange capacity (C.E.C.)	Indicates chemical reactivity of soil by measuring the potential adsorption of exchangeable cations	1.5.3
Total exchangeable bases (T.E.B.)	Indicates availability of exchangeable bases (Ca, Mg, Na, K). T.E.B. = C.E.C. − E.A., reflects basic nature of soil	
Exchangeable sodium percentage	Proportion of C.E.C. occupied by sodium ions; affects permeability	4.8.1
Electrical conductivity of saturation extract	Indicates salinity which may restrict growth	4.8.1
Extractable ions	Indicates nutrient availability; major macronutrients N, P, K especially important	1.6
Toxins	In some substrates, compounds containing B, Zn, Cu, and Mn are phytotoxic	4.9
Loss on ignition	Indicates amount of organic matter	1.5.2
Tissue analysis	Analysis of harvested plants reveals natural uptake	Bradshaw and Chadwick (1980)

PLANT GROWTH EXPERIMENTS
While chemical tests can indicate the potential of the substrate as a growth medium, the real test is to see whether plants will grow and how the material can be altered to improve growth. This can easily be achieved by simple pot or box experiments or setting up small field trials. The design of such experiments is well established and is the province of the specialist.

AFTER-CARE

Site monitoring after the establishment of vegetation is essential. Often this can take the form of a visual inspection and identifies longer-term problems and remedial measures such as further applications of fertiliser.

SITE PREPARATION

Once the major problems have been identified, solutions may be formulated consistent with environmental demands, ecological demands and the ultimate land use. Restoration practice follows the sequence shown in fig. 3.10. Here we shall focus on site preparations prior to planting, aspects of which are developed in section 3.7.

Stability

A precondition for vegetating soils is that the surface should be stable. On slopes, stability refers to deep-seated failures as well as surface mobility. Every civil engineer will be familiar with the criteria for designing stable slopes (see section 3.3.2), but one point worth noting is that many tipped materials are prone to weathering and breakdown. Over a period of time this may lead to a reduction in the cohesive strength, a feature which should be included in stability calculations. Where possible, excess slopes should be avoided and generally no steeper than 1:2 (27°). On steep slopes materials may have to be held in place with mechanical structures. Slopes that are marginal for stabilising with vegetation alone can be held in place by a combination of mechanical and vegetation practices (see Schiechtl, 1980; Gray and Leiser, 1982).

Where the surface layers are prone to water erosion, the basic strategy is to shield the surface using mulching materials and wherever possible to reduce the volume and velocity of surface runoff. The latter can be achieved by flatter cut and fill slopes, flatter gradients in drainage channels and by reducing effective slope length by diversion ditches. Besides the addition of surface mulches, infiltration can be enhanced by applications of calcium as an aid to aggregate formation.

Drainage

Though drainage is an important component of erosion control strategy, it is also essential that water must be available in the right quantities and at the right time, as a buffer against the vagaries of climate, the character of the substrate and the needs of the plant.

When blessed with rainfall all the year round — as in Britain —

water supply is not usually a problem, though moisture deficits occur in the summer months. When water is in excess the area can be drained; this is most easily achieved by controlling the surface runoff using a system of temporary drains (Downing, 1978). The aim should be to keep the water table just below the level of the plant roots.

When the rainfall is of a more seasonal nature a number of problems emerge: first, soil moisture deficits during the dry season will be more pronounced; second, when it rains it *rains*! (in general, the less the annual rainfall, the more intensely it falls); third, intense rainfall on a sparsely vegetated surface will result in a high sediment yield. Thus drainage must be designed to serve both water and soil conservation. Such practices are well documented e.g. Bennett (1955), Schwab *et al*. (1966) and Hudson (1971).

Top soil application

When top soil has been conserved, or if it is readily available – it provides in one operation a fully developed soil with a good structure, texture and an adequate store of nutrients; taken together these should lead to the establishment of a good plant cover. Also, planting can be carried out over a wider period of the year than with raw waste materials. To be effective the layer of top soil must be at least 0.1 m thick, this providing a good seed bed with sufficient nutrients for the early growth phase. Thereafter, plants roots will draw water and nutrients from the underlying material. To be totally self-sufficient, grass requires more than 0.25 m of soil and trees a greater depth. Clearly this is an expensive operation and may not be necessary if the underlying material can be used directly. Further, it does not guarantee that plant growth will not suffer as the roots penetrate the underlying material; if this is toxic or impermeable, the plant will soon exhaust the nutrients in the top soil, or can be vulnerable to drought.

Other materials

Sometimes textural imbalances, or lack of organic matter, can be improved by adding other materials. For example, clay may be added with sandy-skeletal materials, and sand with fine clay soils – intermixing to a depth of 0.10-0.15 m (4-6 in). Organic matter should then be mixed in amounts of 5-20% by volume. This may be obtained from nearby peat bogs, animal manures or digested sewage sludge.

It may also be feasible to exploit other mineral wastes – such as colliery spoils with a low pyrites content. If they are non-toxic and possess nutrient and water retaining properties, they can be most valuable. Usually they pose individual problems, but these may be overcome with careful management (see Bradshaw and Chadwick, 1980, page 81).

Direct improvement

One of the greatest disadvantages of top soil application is its cost. In many cases derelict land materials may be treated effectively and cheaply by normal agricultural techniques; and as such they represent an attractive alternative. Most problems can be overcome, provided they are recognised in advance.

Structural problems, especially arising from excessive consolidation can be overcome by a ripping operation which opens the surface structure.

Problems of excess water are normally overcome by drainage. Where water is in short supply, it may be advantageous for short-term irrigation to be carried out. Water is often present in lower layers and trees in particular have the capacity to reach these moisture levels. Where materials have a coarse structure, e.g. slate waste, this process can be encouraged by designing water-containing pockets (see Bradshaw and Chadwick, 1980, page 208). Incorporated water absorbent materials, such as peat, greatly assist this process.

Surface temperatures are only high in the absence of plants. If plants can be established in cool periods, the problem will be reduced. Transpiration as an evaporative process will automatically reduce surface temperatures. Where there is intense solar radiation, a rapidly growing nursecrop can be used for giving shade to plants which are slow to establish.

Nutrient deficiencies can be overcome by the addition of a suitable fertiliser. In the case of the macronutrients – N, K, P – the fertiliser requirement may be linked to the measured nutrient status of the soil using guides such as the Ministry of Agriculture, Fisheries and Food (1979) and U.S. Environmental Protection Agency (1975).

Beyond this, the interpretation of the chemical analysis of a soil is a job for the expert. Care should be taken to distinguish between the establishment phase and the maintenance needs. In the case of nitrogen, a nitrogen dressing will normally be required during establishment, but subsequently, the needs may be met adequately by nitrogen fixation if legumes are present. The most accurate way of gauging fertiliser needs is likely to stem from an evaluation of the effectiveness of various treatments during field trials.

Toxicity problems are often difficult to overcome and require specific treatment. In the case of acidity, lime can be added in accordance with chemical tests and standard guides such as Ministry of Agriculture, Fisheries and Food (1979). In the case of colliery spoil containing high levels of pyrite, acidity will continue to develop and long-term problems may emerge. Salinity problems may be overcome by exposure to leaching. Otherwise, it may be necessary to select plant species or varieties which are tolerant to the particular conditions.

3.6 Plant materials

3.6.1 Selection and suitability

The most important consideration in choosing plants is that they serve their intended purpose – whether it be erosion control, aesthetics or economic production. For example, if the aim is to prevent surface creep, grasses can be ineffective because of their shallow root system; thus appropriate deep-rooted trees and shrubs must be planted on such sites. In contrast, most adapted grasses can quickly become established on stable slopes and will prevent more surface erosion during the first few years than will trees or shrubs.

Each species has its climatic, physiographic and biologic limitation. Moisture, light, temperature, elevation, aspect, balance of essential nutrients, and plant competition are ecological parameters that favour or restrict all plant species. Each species selected for a particular situation must therefore be adapted to all existing factors of the habitat. Otherwise it is a case of manipulating factors to make them favourable. Adding fertilisers or liming soils are examples of this approach. Manipulation of site conditions will also broaden the range of options for selecting adapted species.

Often the best sources of information for choosing plants are the experiences gained from the revegetation of similar sites (see Hutnik and Davis, 1973). Many books and reports, e.g. U.S. Environmental Protection Agency (1975), Donovan et al. (1976) and Schiechtl (1980) contain lists of plants, summarising features of their habitat, tolerances and value. Native vegetation is often an excellent indicator of the types of plant community which will prevail. Consideration should be given to the availability of seeds or seedlings, resistance to disease and insect damage, fire risk, and so on. The principles of ecological succession are important since the aim is generally to attain a permanent plant cover more than simply to grow plants. Hence the creation of a self-sustaining plant community is an important ecological goal.

3.6.2 Grasses and legumes

A grass mixture containing a legume is one of the commonest means of attaining a vegetation cover because it minimises erosion and helps to condition the soil. There are of course an enormous number of species of grasses and legumes growing throughout the world and it is a matter of selecting the most appropriate. In the context of land restoration it is perhaps their adaptation to soil conditions that matters. Principal factors which should be taken into account are fertility demand, pH tolerance, tolerance to moisture and temperature. In more extreme situations adaptation to saline conditions or tolerance to metals can be important. Correct choice can make all the difference to the permanence and the ease of maintenance of a plant cover.

Trace element deficiencies rarely affect the growth of grass, but if used for herbage, grazing animals can suffer from inappropriate levels of certain trace elements. The trace element content is primarily influenced by soil type and pH.

Legumes are a crucial component of most grass mixtures because they contribute and maintain adequate nitrogen levels. The amount of nitrogen fixed by the rhizobia associated with white clover in pastures is about 100-200 kg ha^{-1} y^{-1} and a typical value for a fertile grassland in a temperate climate is 50 kg ha^{-1} y^{-1} (Bradshaw and Chadwick, 1980). Most productive legumes available require reasonably high levels of calcium and phosphorus for growth and nitrogen fixation.

Grass mixtures often contain a nursecrop. These are rapidly establishing but poorly persisting species, e.g. Italian rye grass, which protect against erosion from an early stage and afford cover for the slower developing long-term sward. Nursecrops should nearly always be sown at low density to allow good light at ground level for the associated plants.

3.6.3 Trees and shrubs

Trees are an excellent means of improving the appearance of the landscape and disguising its scars. With a greater rooting depth they are able to penetrate moisture levels which are inaccessible to shorter-rooted vegetation. Also, by reinforcing soils over greater depths (coupled with controls on soil moisture) trees add to the overall stability of slopes (see section 3.3.2). However, they are less effective than grass in stabilising soil against erosion. For this reason it is strongly recommended that areas designated for trees should be pre-planted with grass. Mixtures of grass seed used for tree-planting areas must be chosen with care to avoid adverse competition effects on the developing trees; high yielding pasture species should be avoided and preference given to slower-growing grasses, e.g. fine-leaved fescues and bent.

In general the most valuable tree species are those which are natural pioneers, e.g. alder and birch. As with grasses, species must be chosen in relation to the particular qualities of the site and climate. Alder has the advantage of being a nitrogen-fixer and there are also a number of leguminous trees, e.g. black locust.

The choice of shrubs is considerable. Since they are unlikely to be used commercially, preference should be given to nitrogen-fixing species, e.g. gorse, and to those which are well-adapted to degraded soils.

3.7 Stabilisation of slopes and earthworks

3.7.1 *Introduction*
Much of the land surface is composed of hill slopes enveloped by a mantle of weathered rock and soil. By comparison with underlying rock, soil is a weak and permeable material which can be readily modified by erosion and human interference. All natural slopes have a tendency to slide when the influence of gravitational and other forces overcome the shearing resistance of the material. A stable regime can be transformed into an unstable state as the result of disturbances in the force balance. External disturbances include seismic activity, soil erosion and man-made activities, such as cutting or filling in part of a slope or ground adjacent to it. Internal disturbances refer to changes in the slope itself. Examples include increases in pore-water pressure – perhaps caused by a rising water table – or factors which cause decreases in the shear strength of slope materials. The human contribution to slope failure is to promote an acceleration of the natural processes causing change or by introducing new factors. Excessive withdrawal of groundwater or intensive mining may cause sudden collapse of the surface. Bare ground can be exposed through overgrazing, deforestation or the worst effects of pollution. Increasing food demand and the shortage of arable land has forced the continued development of hillslopes for cultivation purposes. The advent of monstrous earth-moving equipment has permitted the construction of massive artificial slopes associated with roads, railways and mining operations.

Vegetation is of little value in arresting deep-seated stability problems, since these generally require constructional remedies. For slopes tending to slip, methods include external supports, soil reinforcement or removal of weight by drainage or regrading.

In contrast to deep-seated stability problems, instabilities in the surface can often be remedied by planting vegetation. However a precondition to successful planting is that the surface should be immobilised during plant establishment. This rather begs the long-term objective and calls for a complementary erosion-control strategy. Such measures usually focus on the *bête noir* − namely water movement.

On slopes consisting of cohesionless soils (e.g. coarse soils) or partly cohesive soils, overland flow is likely to cause rill erosion which with time may develop into gully erosion. Here the erosive potential is greatly diminished by reducing the 'active' volume of overland flow and retarding its motion. On long slopes it may be necessary to reduce the effective slope length with a system of diversions or bench terraces which also act as energy dissipators (see Schwab *et al.* 1966 and Schiechtl, 1980). Where slopes comprise cohesive soils (e.g. fine soils), these are not very susceptible to erosion, except in the case of heavily fissured clays where strong water seepage and softening may lead to debris slides. Such situations can be forestalled by instigating drainage measures.

Beyond drainage control, planning depends on the extent of the problem. Where there is considerable surface movement, this can be arrested or prevented by a combination of sediment traps and reinforcements through the mobile depth. In less adverse situations, surface protection methods involving mulching and grassing will often suffice.

Between deep-seated stability problems and surface phenomena there exists a regime in which combined vegetative-structural measures may provide an attractive method for combating erosion and slope failures (see Gray and Leiser, 1982; Schiechtl, 1980).

The key to success for vegetating low-productivity soils or in difficult terrain, is by good planning (see section 3.5.1). Though there is a vast literature on this subject, readers are referred to U.S. Environmental Protection Agency (1975), Schiechtl (1980), Gray and Leiser (1982) and Hackett (1972), which between them provide, a balanced account of the planning, design and constructional phases together with case-book examples. Where engineering works are required these should be consistent with codes of practice such as the British Standard BS 6031.

3.7.2 *Drainage*

Basic drainage requirements were outlined in section 3.5.3, but deserve amplification in the context of slope stability. This is greatly enhanced when measures are taken to prevent the ingress of water into the slope and preferably to remove water already in the slope – though not to the detriment of the basic needs of plants.

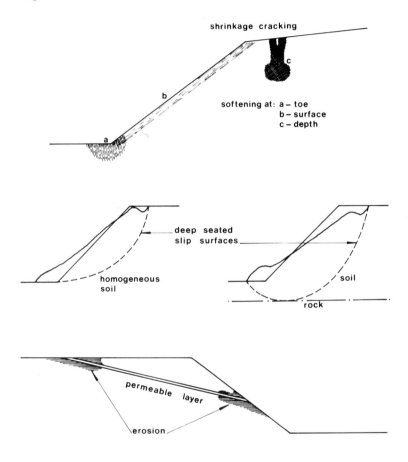

Fig. 3.11 Critical zones in slopes; shaded areas indicate danger zones to be protected by suitable drainage facilities

Figure 3.11 illustrates a number of critical zones in a cut slope which must be protected by suitable drainage facilities. Drains are normally constructed along the top of a cutting to intercept surface and near-surface water and so prevent it from flowing down the slope. Within the slope, buttress drains are often used as remedy for slips in cuttings, removing water from within the slope and providing weighty buttresses of stone or rocky material capable of resisting any tendency to slip. They also act as collectors for shallow auxiliary drains which fan out over the slope face to tap springs and seepage flows and to intercept water flowing down the face of the cutting. Drainage is also required at the slope base to collect intercepted water from above and to maintain the strength of the toe – *a site of potential instability*.

In conventional engineering practice, drains are usually constructed using inert materials. There are however, situations in which techniques employing vegetation have been used to advantage as a low cost option.

Grassed waterways have often been used in road construction and are widely used in agriculture as a means of draining and channelling surface water on slopes, and particularly to drain surface water around the top of a slope. A grassed waterway alone is seldom safe when applied to slopes $> 11°$, though with masonry drops they can be used for draining water from terrace systems on slopes of up to 20° (Sheng, 1977).

For steep slopes, surface waters can be transported downslope using a system of 'live fascine drains' which are described in Schiechtl (1980). Basically, these are ditches running downslope into which are embedded fascine rolls (a 0.2-0.4 m diameter sausage-like bundle of live branches tied together with wire) and covered with soil. The fascine drains are effective in transmitting water immediately after placement, due to the channelling effect of the longitudinal branches. After the plants have formed roots they have further potential for soil drying, resulting from transpiration.

3.7.3 Soil stabilisation methods using live materials

Soil stabilisation methods are designed to provide fairly rapid control of major surface movement. This is achieved using a point by point system or successive lines of soil reinforcement across the slope exploiting the rooting systems of woody plants. Soil stabilisation methods are usually supplemented by a surface protection system which affects a larger area.

Principal methods include submerged wattle fences together with cordon and brush-layer constructions which are set out terraced fashion. As an example, a brush-layer construction is described below.

BRUSH-LAYER CONSTRUCTION

Commencing from the bottom of a slope, trenches or terraces are cut, into which are placed live branch cuttings as depicted in fig. 3.12.

A lower ditch or berm is then filled with spoil from above. The branches have the advantage of trapping soil washed down the slope and reducing the velocity of runoff. This effect, together with soil reinforcement, starts immediately upon construction and increases with root formation.

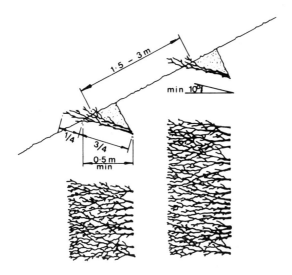

Fig. 3.12 Diagram of a brush-layer construction on a cut slope; work proceeds from bottom to top

Brush-layers and the closely related 'hedge brush-layer construction' have been found to be very effective for the rapid stabilisation of cut and fill slopes in extreme sites, and the method can also be used on river banks. Together they rank among the most important techniques for use in rocky, loose material. A disadvantage is that they are unsuitable for retaining top soil except when longitudinal strips of an inert material are introduced and placed over the lower edge of the berm to protect the edge against erosion.

3.7.4 Surface protection methods
The principal function of surface protection methods is to provide protection against surface erosion over large areas where the depth effect in the soil is of minor importance. Methods include brush-mattress construction (see Schiechtl, 1980), turfing, standard seeding, hydroseeding, mulch-seeding and others. Seed establishment can be greatly assisted by the use of mulches and stabilisers.

TURFING

In ordinary circumstances, any advantages which might accrue from the use of turf will be outweighed by economic and practical considerations. It is expensive to obtain and lay, and unless selected by an expert, may not contain suitable species. The best arrangement of turves is to cover the area completely, but this is rarely possible because there are not enough sods available. This can be remedied by planting in a checkerboard pattern or strips in conjunction with top-soiling and seeding in the intervening spaces. On steep slopes it may be necessary to secure the turves with wooden or steel pegs.

HYDROSEEDING

One of the major problems on steep slopes and on derelict land is in the initial establishment. Hydroseeding or hydromulching is a technique in which seeds and nutrients are sprayed over the ground as a slurry.

In normal agricultural practice, seeds are incorporated into the soil which is then consolidated by harrowing or rolling. With difficult terrain and problems of soil texture, these well-established sowing techniques cannot be used. Hydroseeding is a method which attempts to circumvent some of the problems.

By suspending the seed in an aqueous mixture it can be sprayed. Good consistency is secured by adding agents such as alignates, starch derivatives, latex and oil-based emulsions which make the mixture more viscous. They also cause the mixture to adhere better to the ground when it lands. In addition the latex and oil-based emulsions act as soil stabilisers. To reduce the translocation of seeds and to prevent dessication, high bulk materials, e.g. straw, peat and wood-pulp are added to the mixture; these provide minute shaded and moistened sites where the seed can germinate and seedlings develop. Nutrients are usually added to overcome potential deficiencies and lime is included to correct acidity problems. Legumes are sometimes included in the seed mix to cater for long-term nitrogen supply. However this can pose problems as legumes are slow to germinate (increasing their vulnerability) and at the seedling stage are very sensitive to solutions of nitrogen and phosphorus which are present in fertilisers. Bradshaw and Chadwick (1980) remark that if fertilisers are left out of the original mix, legume germination is excellent. The mineral fertilisers can then be added later.

Ideally the applied layer forms a resistive crust and provides an optimum environment for germination. The amount of the material required and its composition depend on local circumstances.

Hydroseeding has been widely used for vegetating steep rocky or gravelly slopes or large areas such as the shoulders of motorways. Details of the system are summarised in the German standard DIN 18 918.

MULCHES

Mulching is a very common method for promoting temporary protection of exposed erodible surfaces to permit the establishment of vegetation. Mulches are effective because when spread over the surface they protect it from splash erosion, retard runoff, trap sediment and create a more amenable microclimate to assist germination and the early development of plants. For instance, a mulch moderates temperature changes, reduces evaporation and conserves soil moisture.

A soft and protective mulch may be provided by almost any organic or inorganic material that is not toxic to plants. The most commonly used natural mulches are straw, and hay. Other natural mulches include leaves, peat, sawdust, wood chips, bark residue, manure, gravel and stones. Synthetic materials include organic and inorganic liquids that can be sprayed to form a thin film, e.g. latex and asphalt emulsions. One may also include a variety of plastic materials such as polythene and polyvinyl chloride. Typical application rates and related information are summarised in table 3.9.

Synthetic mulches

Synthetic mulches are sometimes termed 'stabilisers' because of their adhesive qualities, allowing the binding of soil particles or natural materials which are prone to being blown away. Where a film is formed, it does not penetrate deeply into the soil and is not readily destroyed by wind or rain. Asphalt films for example, are virtually non-porous; this can be a disadvantage as much of the rainwater runs off. Ideally the surface film should be stable against erosion, sufficiently porous to allow water to enter, yet sufficiently persistent to allow permanent vegetation to become established. Stabilisers of the bitumen or latex type are widely advocated for sand stabilisation.

Conventional mulches

Unlike synthetic mulches which cling to the soil, the key mulching properties of conventional mulches arise from the formation of a barrier of still air close to the soil, together with a capacity to absorb water and the momentum of raindrops. Through decomposition, they add the all-important organic matter to the soil, forming a basis for the natural development of humus.

Straw or hay generally has to be treated with a binder or anchored to the ground to stop it blowing away or bunching under the action of flowing water. Excellent results have been obtained by covering the straw with jute netting which may also be used alone (Hackett, 1972 and Dallaire, 1976).

Table 3.9 Surface stabilisation can be important: mulches and stabilisers suitable for derelict land reclamation

Material	Rate (tonnes/ha)*	Persistence	Stabilisation	Soil water retention	Nutrient	Toxicity
Mulches						
Excelsior	4	OO	OO	OO	o	o
Wood shavings	4	OO	O	O	o	o
Wood chips	10	OO	O	OO	o	o
Bark shredded	4	OOO	O	OO	o	o
Peat moss	2	O	O	O	o	o
Jute netting	–	OO	OOO	O	o	o
Corncobs	10	OOO	O	O	o	o
Hay	3	O	O	O	O	o
Straw	3	OO	O	O	o	o
Fibreglass	1	OOO	OO	O	o	o
Stabiliser/mulches						
Wood cellulose fibre (as slurry)	1 – 2	OO	OOO	O	o	o
Sewage sludge (as slurry)	2 – 4	O	OO	O	O	o
Stabilisers						
Asphalt (as 1:1 emulsion)	0.75	O	OO	O	o	O
Latex (as appropriate emulsion)	0.2	O	OO	o	o	OO
Alginate or other colloidal carbohydrate (as emulsion)	0.2	OO	OO	o	o	o
Polyvinyl acetate (as 1:5 emulsion)	1	OO	OO	o	o	OO
Styrene butiadene (as 1:20 emulsion)	0.5	OO	OO	o	o	OO

OOO high   O low   o nil

*Rates can be varied depending on circumstances – will affect soil water capture and retention and seedling establishment

Mulch-seeding

Mulch-seeding is a development of normal mulching, whereby a seed mixture and fertilisers are incorporated into the mulch.

In the Schiechteln method described in Schiechtl (1980), the mulching procedure is split into three stages: in the first stage, long uncut straw is placed on the slope so that an unbroken, complete mulch layer is formed; in the second stage, the seed mix and fertilisers are spread onto the straw; and finally, the cover is protected against movement by spraying it with binders. Where this is inadequate the system can be immobilised using spikes and wiring or secured by mesh.

The Schiechteln method has developed into the most widely used revegetation method in alpine and Mediterranean areas. The variation using spikes and wire to hold down the mulch makes it possible to revegetate rocky or very smooth steep areas, e.g. with clayey soils from which the material would slide away if other seeding systems had been used.

3.8 Stream bank protection

Farmers and communities have a vested interest in bank protection and river management, not only from the flood hazard, but also because areas destroyed by stream bank erosion usually represent high-value agricultural land. Alluvial channels rarely follow a straight course and often have a meandering or braided pattern. Meanders tend to migrate even when the river system as a whole is dynamically stable. Rather than considering natural erosion as an evil, it should be viewed as a process in which a river finds an equilibrium with its environment. In recent years, however, recreational and other uses of rivers have increased and tend to exacerbate erosion rates. In larger bodies of water, e.g. reservoirs, water is virtually stagnant and wind-generated waves may be the principal erosive agent. Whatever the cause, it is crucial that the engineer should have a good grasp of the problems that are being tackled, lest solutions in one place create problems elsewhere.

Many types of engineering works may be constructed to prevent bank erosion. The choice is often difficult because of the interaction between technical, economic and conservation factors. Most often, engineers will look towards techniques based on inert situations. In many instances these are essential. However, it is likely that bio-technical methods — whether based on vegetation alone or in con-junction with inert materials — can be applied in all other situations.

The Bavarian state authorities have been in the forefront of developing biotechnical methods in the context of river engineering;

their work is reported in Binder (1979), Pfeffer (1978), Miers (1977) and elsewhere. The philosophy of their approach is described in Seibert (1968) and Bache and MacAskill (1981), with supporting practical detail described in Schiechtl (1980). Conventional practices based on inert materials are reviewed in Keown *et al.* (1977) and Charlton (1980).

3.8.1 *Engineering perspective*

FLOW MAINTENANCE

When planning river works, the salient feature to recognise is that streams and rivers are important components of land drainage, whose primary function is the removal of surplus water. This embraces the control of the water table in agricultural areas for optimum growth conditions, and similarly, the disposal of surface runoff and effluents without the creation of flooding. This implies that flow must be maintained and, in general, water authorities possess considerable power to ensure this basic requirement. The general approach to planning is described by the Institution of Water Engineers (1969), the U.S. Department of Agriculture (1975), and a broad view embracing conservation interests is given by the Water Space Amenity Commission (1980).

River authorities may object to plants within and around water courses for a number of reasons. First, because shrubs and trees can hinder the normal maintenance of a water course; and second, because plants and plant debris can obstruct the flow of water and create flood hazards by causing increases in the flow depth. Increases in flow depth can be generated in a number of ways. First, the submerged volume of vegetation causes an upward displacement, but more significantly it causes the stream velocity to be reduced. When this occurs the required cross-sectional area, and hence the depth to accommodate the flow, must increase. In addition, a consequence of reduced flow is an increase in the probability of the river shedding its sediment load and promoting accretion, thereby further reducing the effective depth of the channel.

The effect of vegetation on channel flows can be roughly determined by selection of the appropriate roughness coefficient for inclusion in the Manning formula (section 2.6.1). Although values of Manning's n can be taken from published tables (e.g. tables 2.21 and 2.22), Institution of Water Engineering (1969) recommends that it is preferable to be guided by the coefficient applying to the channel before improvement unless improvement is very drastic. For small channels (typically encountered in land drainage) the approach discussed in Miers (1974) allows for seasonal variations in plant growth.

REGIME

An important feature of river morphology is that channel patterns are not entirely random, but conform to 'rules' (albeit empirical) which reflect equilibria between the river and its environment. For example in gravel rivers, when sediment transport rates are small, Charlton (1978) cites:

$$b = 3.75 \, k \, Q^{0.45} \qquad\qquad [3.16]$$
$$1.3 > k > 0.9 \text{ (grass and light vegetation)}$$
$$1.1 > k > 0.7 \text{ (trees and heavy vegetation)}$$

where b represents the channel width (m), Q the bank-full discharge ($m^3 \, s^{-1}$) and k, a coefficient dependent on the vegetation bordering the river. Similar correlations may be found between other variables including the effective depth, the size of bed material, the hydraulic gradient, the amount of sediment transport, and so on. The significance of such correlations is that they reflect a hydraulic regime which is 'stable' and should be considered in channel design. A useful case-book example is described in Miller and Borland (1963) in which local attempts to stabilise a channel by bank shaping and planting vegetation were thwarted, with erosion continuing at an even more rapid pace as the project proceeded. Eventually a holistic approach to channel stability was adopted, taking account of regime principles. Revised control measures were implemented and met with conspicuous success.

CAUSES OF BANK EROSION

Banks recede as a result of abrasion and scour, or arise from bank collapse. Abrasion is caused by high flows, wave action or animals and people trampling up and down the bank slope. Bank collapse is often induced by the seepage of water following rapid drops in water level, loss of strength due to increased moisture content, or strength losses caused by the sequence of drying followed by shrinkage and cracking. Fluvial factors connected with alterations in discharge rates and sediment may also generate conditions which cause bank erosion.

METHODS OF BANK PROTECTION

Banks can be protected by controlling abrasion or enhancing their internal stability. The latter is discussed in section 3.7.2, salient features including the control of the soil moisture regime and increasing the soil strength (see Charlton, 1980). Abrasion control can be considered under the headings of bank armouring, flow retardation and flow deflection.

In each of the above the value of the vegetation for protecting the soil substrate depends on the combined effects of roots and foliage. Roots and rhizomes provide reinforcement — increasing the shear strength of the substrate. Immersed foliage elements absorb and dissipate flow momentum and may promote sufficient attenuation to prevent scour. Although the buffering capability is advantageous, foliar abundance can also promote problems due to drag, e.g. during times of spate, large bushy plants with poor anchorage are liable to be torn from the banks. Haslam (1978) cites trees as the most effective for preventing erosion followed by shorter grasses. This feature is reflected in equation [3.16] which shows that the type vegetation on the banks affects the width of river, with tree-lined channels generally narrower. The survey by Charlton (1978) indicates that the correlation between width and discharge for the lightest conditions of bank vegetation is almost the same for sand-bed channels when the channel is free to develop its width; this suggests that light vegetation bordering gravel rivers is not particularly effective in reducing erosion and emphasises the importance of trees for increasing the resistance of river banks.

Bank armouring: revetments
This is generally the most versatile of all methods and will nearly always prevent erosion due to abrasion and scour irrespective of river gradient or velocity. Revetments are limited only by the strength and weight of the construction materials. Piles, concrete, brickwork, blocking, rip-rap, gabions, and so on, are used in situations where the preservation of the banks is vital and where biotechnical methods are inappropriate. Biotechnical equivalents include branch mattresses, live branch packings, grass sods and tree plantings (making use of root systems). Many of these methods depend on combinations of live and inert materials. At this time there is little information showing the conditions that can be tolerated by live materials, though these are usually confined to situations with slow flows and fairly stable water levels. In the case of grass sods, these can tolerate velocities of up to ~ 1 m s^{-1} for long periods and higher velocities for shorter times (see section 3.11.1). When the data leading to these criteria is translated into critical shear stresses, Parsons (1963) showed that the resistance of a good sod has an equivalent stone

diameter of 50-130 mm. This equivalence is useful since structures like rock rip-raps are usually designed with stone sizes in mind (see U.S. Department of Agriculture, 1975, chapter 16). In the case of more sturdy biotechnical constructions based on combinations of woody vegetation and rocks, one should expect a higher range of tolerance.

In deep channels or with higher flows (say with stream velocity $\gtrsim 1$ m s^{-1}) the only effective and practical means of ensuring the integrity of the bank is by rock revetment or other structures. Experience suggests that vegetation is effective in controlling water above the sustained high-water level (Mifkovic and Petersen, 1975) and can be useful in controlling water down to the normal water level.

Flow retardation
Since abrasion varies as the flow velocity cubed (roughly) above a critical value, there is considerable merit in retarding flows. When carried out near to banks, this has the advantage of promoting deposition and accretion. However, when considering the channel as a whole, flow adjustments must be consistent with the transport of sediment load, if overall stability is to be preserved. Where the flow entering a section to be protected carries only silt and fine sand in suspension, the maximum velocity should be limited to that which is non-scouring on material of the smallest size occurring in any appreciable quantity in the stream bed material; the minimum velocity should not be less than that required to transport the suspended material.

Reductions in flow can be achieved by the introduction of weirs, sluices for raising upstream water levels, together with cut-offs which act as by-passes to flow. Local retardation is attained using training fences and jacks. Training fences are suitable in streams with low gradient and high sediment load which can be deposited behind the fence, so building up the bank. These are particularly useful for repairing eroded embankments (see figs. 3.13 and 3.14). Biotechnical methods include live ground sills, brush and comb constructions, brush mattresses, brush-layer barriers, log-brush barriers, live fascines and reed buffer zones, each of which are described in Schiechtl (1980).

Flow deflection
Spurs are a useful means of diverting flows and can be more economic than revetments. They can be oriented upstream or downstream deflecting the water towards or away from the zone of interest.

Generally, in gravel rivers, spurs are oriented at right angles to the flow or inclined slightly downstream. When used in groups, Charlton (1980) quotes a useful criterion dependent on flow depth and channel roughness for determining the spacing between spurs. In the vertical plane, sills can be used for hindering scour and have the advantage of raising the channel bed downstream.

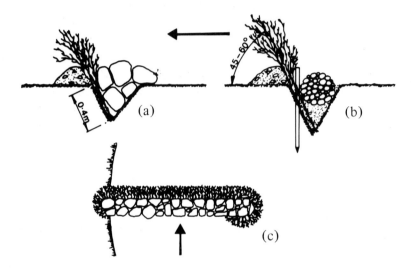

Fig. 3.13 Live siltation construction: (a) live siltation construction with rip-rap; (b) with fascines for stabilisation; (c) plan view

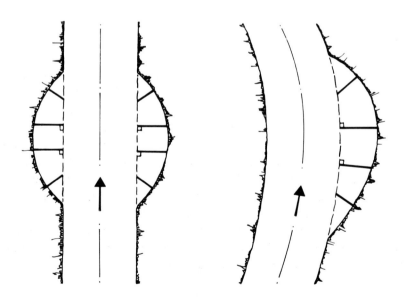

Fig. 3.14 Schematic arrangement of a live siltation construction for restoration of a shore break

3.8.2 Vegetation along watercourses

River plants are sensitive indicators of their environment; thus if they are to be exploited as a medium, it is vital that plantings should be consistent with the natural environment of the plant. Key factors controlling distribution include flow, substrate, light and nutrient availability. Of these, Haslam (1978) stressed the importance of water movement — describing the flow pattern as an overall controlling factor 'comparable to the action of drought in the desert or cold and dark in the Arctic.' In the same vein, Seibert (1968) lists flow, level and movement as the factors which primarily determine both zonation and the type of plant growth able to colonise any given part of the flow cross-section. Figure 3.15 illustrates the correlation between species zonation and the duration of water levels, and represents the complete series of plant communities around continental waters. In the simplified scheme described in Seibert (1968) four zones are identifiable:

(a) zone of aquatic plants, permanently submerged;
(b) reed-bank zone, the lower part permanently submerged for about half the year only;
(c) zone A, flooded only in periods of high water, and colonised by woody plants, i.e. shrubs and trees;
(d) zone B, flooded only in periods of very high water, and dominated by trees.

Seibert reviewed the plant associations in each zone and their value and implementation in bank protection.

AQUATIC PLANT ZONE

The plants of these communities generally require a slow current and sufficient light, their growth being inhibited by the shade of trees growing along the banks of narrow streams. While they may protect the bed against denudation, it is important that they do not become too luxuriant, since the flow will become restricted. In lowland reaches of rivers Haslam considered that vegetation is safe if the river is 25% full (i.e. when looking at the stream it appears that 25% of the water volume is occupied by plants). Greater amounts of vegetation can grow to a dangerous level during a spell of warm weather. This contrasts with mountain and upland streams, where a much higher percentage of submerged vegetation can be tolerated. Though such observations may be useful, it is essential to evaluate these aspects using Manning's equation or its equivalent (section 2.6).

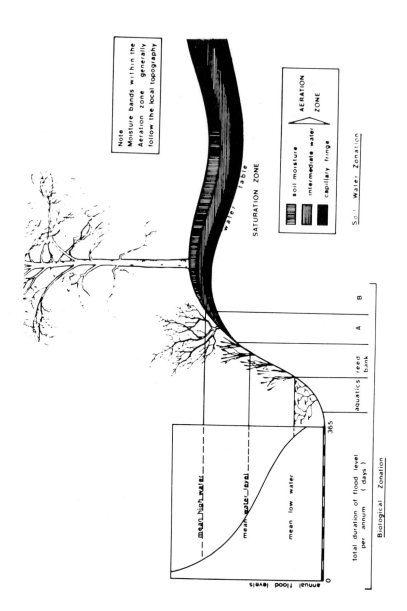

Fig. 3.15 Correlation between vegetation zonation, water-level duration, and soil water

The artificial implantation of aquatic plants is difficult and has seldom been tried (Seibert, 1968) and river authorities try to discourage their presence.

REED-BANK ZONE

Reed-banks are often used in canals and rivers as a means of protecting the shore or margins against erosion caused by water flow and wave attack. The protective capability arises from soil binding and by virtue of their submerged shoots which dissipate flow momentum. According to Schiechtl (1980) reeds provide better protection against waves than any other plant. Reduced water velocities also enhance the opportunity for sedimentation and accretion. Their protective value varies widely and Seibert (1968) states that the greatest degree of protection is provided by the common reed and is attributable to its robust nature and strong anchorage.

Depending on ecological conditions, the reed-bank zone may contain very different associations of reeds and sedges which generally grow in stagnant and slow-flowing water at depths up to 2 m. Many species, however, are confined to depths of < 0.3 m in clay-silt to fine sandy soils (Seibert, 1968).

Miers (1977) indicates that reed-banks have been successfully used where the velocity does not exceed 1 m s^{-1} and that their effectiveness depends on their extent and the incident flow. The width of the bank is determined by the inclination of the shore slope and the depths within which the reed will thrive; for example, Pfeffer (1978) quotes that a slope of 1:3 produces a plant strip only about 1 m wide, half above the water level and half submerged. In faster-flowing waters or in navigable rivers, plants by themselves may be ineffective (Miers, 1977) and it is usually necessary to reinforce the shore from the bottom to the middle water level using an inert material, but with adequate spaces left for plants if this is desired (Pfeffer, 1978).

Bonham (1980) examined the effectiveness of reed-banks in cushioning the effects of waves caused by motor boat wash. For the species studied and under suitable conditions of depth, vegetation density, and a bed slope of 1:4, a bank 2 metres wide was able to dissipate almost two-thirds of the wave energy.

Implantation

There are various ways of planting reeds and sedges. Planting in clumps is the oldest method and suits plants of this type. Because of the weight of the material involved, transportation costs may be high and there is also a limited availability of material. Less material is required for the planting of rhizomes and shoots. The most common and economical method of establishing reed beds is the planting of culm slips. Like other implantation methods it takes time (2-3 years) for a closed dense reed belt to become established and to attain its maximum protective capability.

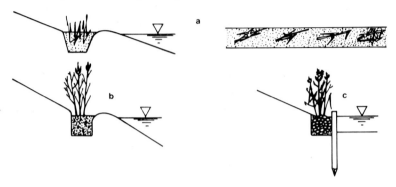

Fig. 3.16 Planting of: (a) rhizomes or shoots; (b) of clumps; (c) or reed rolls, to protect the river bank

In many cases these methods do not consolidate the banks sufficiently during the period immediately after construction. Combined structures have therefore been designed in which protection of the bank is at first ensured by inanimate materials. The most effective of these has been the use of reed rolls as illustrated in fig. 3.16.

WOODLAND ZONE

Haslam (1978) describes trees as the most old-fashioned, cheapest and often most useful means of decreasing river plants (by shading) and stabilising river banks.

Zone A occurs predominantly in areas prone to seasonal flooding and on low-lying soil. In Britain this is often referred to as 'carr' or 'marsh woodland'. Seibert and others have somewhat ambiguously described zone A as the 'softwood' zone, recognising the physical nature of the wood rather than the generally accepted definition in forestry practice in which softwoods refer to conifers. It is characterised by the fast-growing species of willows and alder, though on gravelly soil the black poplar may also be present. Alderwood (described in Tansley, 1968) is the natural climax of marsh woodlands. Zone B is referred to by Seibert as the 'hardwood' zone, and is less significant for bank protection because it is less often flooded and less easily eroded by water. The zone comprises associations of ash-elm, alder-ash, and others described in Seibert.

Seibert stressed that stands of full-grown trees are of little use for protecting banks, and are most effective when treated as shrubwood. In terms of species, Kite (1980) considered that alder is most effective for long-term bank protection because the roots are water-tolerant and support the bank material by penetrating down to a considerable depth within the aeration zone. Under water, the exposed and densely-matted root system is an important asset in the protective mechanism. Kite also notes that the roots of the trees from the B zone, such as ash and sycamore, tend to exhibit lateral growths along the bank and above the water level and suggests that this can exacerbate undermining.

According to Seibert, species in zone A readily throw up suckers, and willows are also capable of developing secondary roots on cut limbs and of bringing out many dormant buds, an attribute which makes them particularly useful for implantation techniques. In contrast, alder branches do not throw out shoots and, with the arborescent and shrub species, they have to be planted in the usual way.

Along small streams bordered by alders, and also medium-sized watercourses with low banks and approximately constant water levels, the root system can be relied upon to preserve the bank almost vertical. However, banks are usually graded to 1:2 or 1:3, with trees planted just above the mean water level in summer. Seibert has stated that this distance should not be more than 0.5-1.0 m if protection in this zone is to be achieved, and a greater spacing serves little purpose unless a belt of perennial herb colonies or turf is maintained in the intervening space.

Pfeffer (1978) suggested that plantings parallel to the course of the current cause least interference to the normal flow. However, this approach is questionable in terms of local erosion. There is also merit in breaking the mass flow by more scattered planting and aesthetically this may be less monotonous. If the shading potential of trees is to be exploited as a means of channel weed control a number of ecological and practical problems emerge which are discussed further in Kite (1980) and Haslam (1978). Choice of trees species (see DIN 19 657) depends very much on natural conditions: a guiding philosophy is to generate a stand approaching the natural state taking account of soil moisture, soil pH aspect, etc. Miers (1977) describes a favoured planting scheme with willows between the stones plus alders to succeed the willows. Toward higher ground, there should be a transition to the species of zone B.

Implantation of riverside woods

In many biotechnical engineering constructions dependent on species from zone A (usually willow), implantation is achieved either individually as cuttings, slips or stems, or bound together in various ways in order to ensure immediate protection of the bank. They may be wattled or wired together as fascine rolls.

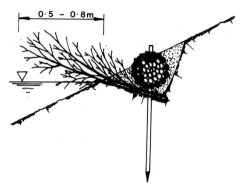

Fig. 3.17 Design of a live shore fascine

Fascines protect the banks and shore immediately after installation, especially if they have been placed on a brush-layer (fig. 3.17). The branches of the brush-layer give protection against waves and washout. Given time the parts of the system in contact with the ground take root. Covering with earth improves this contact and retards desiccation.

There are many different structures which can be made from combinations of live branches and stones, as exemplified in fig. 3.13.

3.9 Reservoirs

In the past the public have tended to be excluded from reservoirs in the U.K. but reservoirs are now considered in terms of their amenity value – providing sporting and recreational facilities as well as offering aesthetic enjoyment. Unfortunately in the summer, when recreational use is at its peak, water levels tend to be low (subject to the vagaries of the British climate!) – exposing bleak margins devoid of vegetation. Most species are unable to colonise this zone because of their susceptibility to flooding and wave action. Persistent wave attack can also cause erosion problems unless remedial measures are pursued. Revetments constructed from inert materials represent an obvious option; while they may be satisfactory from the engineering viewpoint, they are expensive and not particularly attractive. In some cases, particularly with older reservoirs, nature has done its best and found plants which will colonise the margins. In more protected areas with a water level variation of $\lesssim 1$ m, herbaceous material can survive. In more extreme conditions, however, the main colonists are shrubs, common sallow and goat willow in Europe and yellow willow in North America. These observations suggest that there is potential for implantation in the margins of new reservoirs; when established, these mollify the appearance of the margin at times of draw-down, and also offer scope for margin stabilisation.

Little and Jones (1979) examined the uses of herbaceous vegetation in the draw-down zone. For U.K. conditions, their study provides specific guidance on the selection of species and their tolerance to varying hydraulic conditions − including their erosion control potential. In spite of this progress, there is still an acute lack of quantitative data for the design of strategic buffer zones, though useful experience has been gained elsewhere: Bonham (1980) examined the performance of reed-banks in canals for absorbing wave energy generated by boat wash; Newcombe et al. (1979) and Department of the Army (1978) provide a strategy for constructing marsh-banks as a coastal defence measure.

The use of trees for margin stabilisation has attracted some interest, with documented examples cited in Gill (1977). As Gill and Bradshaw (1971) remark, there is no use in planting trees on margins unless the ecological problems are understood. In waterlogged conditions, tree growth is dependent on satisfactory root growth, the latter depending on adequate oxygen supplies − particularly during the period of active growth. Any tree can survive short periods of flooding, but their flooding-tolerance differs markedly with species (see Gill, 1977). In general, adapted trees can probably survive prolonged periods of partial inundation, provided that they are unflooded for at least 60% of the growing season on average. Therefore a prerequisite to tree selection and planning is a knowledge of the frequency distribution of water levels and flooding. Observations of natural tree growth around lakes and reservoirs show that the summer median water level is critical; in no case can much tree growth be found much below this; plantings can generally take place down to about 2 metres below the top water level (Gill and Bradshaw, 1971). Species also differ in their relative resistance to wave action and undermining. (See Gill and Bradshaw who also discuss the significance of soil type, the impact of grazing animals, and the organisation of trial plantings.)

3.10 Gully erosion

Gully erosion is spectacular and widespread. In some countries it is a serious problem, gullies encroaching on agricultural land, threatening roads and property and endangering livestock. Materials eroded from gullies reduce the capacity of reservoirs and cause excessive silting in rivers and streams. They occur more frequently in semi-arid areas, with infertile soils and sparse vegetation.

3.10.1 Processes in gully erosion

Gullies are caused by runoff water flowing at a velocity sufficient to detach and carry away soil particles. Behind this, there exists a factor which is common to all cases of gully erosion. In engineering terms, the cause is a breakdown of metastable equilibrium in the stream or watercourse. A watercourse is usually in regime, i.e. the size of channel, its shape and gradient are in equilibrium with the flow it carries. If the balance is disturbed, say resulting from a larger flow, a stream will tend to come back to equilibrium with a modified cross-section or gradient. However if the disturbance is too great, the stream may start gullying, such that the combination of hydraulic radius, roughness, coefficient, and stream gradient results in a continued state of increased velocity (see Hudson, 1971, page 216). This is one reason why gully erosion is nearly always self-perpetuating and not self-correcting. Disturbances can take the form of increased flows or a decrease in the ability of the channel to carry the flow. Increased flows can be generated through changes in land use or as a result of poor land management, e.g. overgrazing, where the infiltration capacity is reduced. Decreases in channel capacity may arise from alterations in channel shape or from increases in channel roughness — perhaps caused by growing vegetation. If flood waters cannot flow quickly enough, overbank flow may occur and start new erosion patterns.

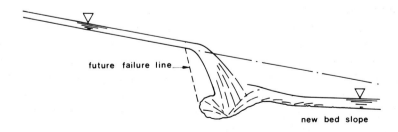

Fig. 3.18 Schematic view of head-cutting in a gully

The two main processes in gully erosion are down-cutting and head-cutting. Down-cutting of the channel bottom leads to gully deepening and widening. Head-cutting is illustrated in fig. 3.18, in which it is seen that the head of the gully works back upstream by a process of scouring, undercutting and bank collapse – this is usually the most actively eroding part of a gully.

3.10.2 Principles of gully control

The first principle of gully control is to determine the cause of gully erosion and to take counter-measures. The second principle is to either restore the original hydraulic balance or to create new conditions. Where excess flows are the root cause of the problem, measures may be taken to direct flows or to increase infiltration and temporary storage within the surrounding catchment. Since such measures may take a considerable period to implement, they should be complemented by treatment of the gully itself. This can generally be attained more quickly because of the concentration of treatment and the availability of higher soil moisture in the defined channel. Short of filling in the gully and modifying catchment drainage patterns, the long-term objective of most gully control work is to stabilise the gully surfaces by vegetative means.

Unfortunately, there are considerable obstacles to the establishment of vegetation in gullies. The environment is usually as inhospitable as it can be, the soils lacking stability, structure, organic matter, available nutrients and a low moisture-retention capacity.

If growing conditions do not permit the direct establishment of vegetation, engineering measures may be required. For example, only rarely can vegetation alone stabilise head-cuts because of the concentrated forces at this location. If the gully gradient can be stabilised, vegetation can become established on the bed. This aids the stabilisation of the banks since the toe of the gully sideslopes is at rest.

The most commonly applied engineering measure for stabilising the gradient is the utilisation of check dams. The purpose of these is to slow down the water so that it travels at a non-erosive velocity and at the same time induces siltation. They are most effective when designed as porous structures since this reduces the forces against the dam and surrounding banks and also aids the dissipation of energy. The spacing of such dams depends on their height and the gully gradient. Generally the steeper the gradient the closer the spacing, and specific guidelines are provided in Gray and Leiser (1982) and Heede (1977); the latter study also describes systems for head-cut control.

Bank sloping should be carried out only to the extent required for the establishment of vegetation or tillage operations. Schwab *et al.* (1966) suggest that where trees are to be established, rough sloping of the banks to about 1:1 should suffice. Where gullies are to be reclaimed as grassed waterways, sloping of banks to 1:4 or flatter is often desirable.

3.10.3 Revegetation of gullies

If runoff which has caused the gully is diverted and livestock excluded from the gullied area, a natural plant succession will develop. This process is slow (see section 1.7) but can be aided by adding fertilisers and mulches. Natural revegetation is certainly the cheapest method of gully control but can only be applied to situations where the erosion is not critical.

When artificial revegetation is considered, its choice should be consistent with the intended use of the reclaimed gully. A stabilised gully may be used as a grassed waterway, a wildlife habitat, a woodland area or pasture. General requirements for plants to be suitable in gully control are that they should grow vigorously in poor conditions and give a good ground cover. A spreading, creeping habit is much better than an upright habit and plants should be deep-rooted. Long flexible plants, e.g. tall grasses are liable to be flattened by the impact of flows and give good bottom protection. The smoothed interface, however, reduces the surface roughness and the resultant higher flow velocities may endanger meandering gully banks. Trees grown beyond the sapling stage may restrict flows and cause diversions against the banks — perhaps causing damage.

Small gullies can be secured by placing dead branches or bundles of branches in them, butt ends pointing up the gully, and then placing large rocks on top; these retain the bedload and boulders, and promote siltation. When all the branches are completely covered a new layer of branches can be placed on top and the process repeated until the gully is filled. This system can only be used in very steep narrow gullies and has been successfully applied in the European Alps to secure steep slopes against potentially dangerous washouts. For repairing shallow gullies (up to 3 m depth and 8 m in width) gully layering with live branches is an appropriate technique provided the bedload and boulder movement are not too severe, and that slow siltation takes place; this enables the soil mass to become firmly anchored and stabilised, and it can build up without the plants suffocating. Each of these techniques is described in Schiechtl (1980).

An alternative approach suitable for arresting small gullies in watersheds of small to medium size is to introduce sod-strip checks along the gully bottom (see Bennett, 1955). The strips should be at least 0.3 m wide, lying flush with or slightly below the bed of the gully, and extending up the gully sides to just above the level of expected high water. Their spacing will depend on the drainage area and the gully gradient, but intervals of 1.6-2 m are commonly specified. The entire gully should also be seeded. A similar technique described in Schiechtl (1980) relies on live ground sills to act as checks and to induce siltation. Given time, this should lead to a gradual filling in of the gully.

It is seldom possible to predict which plants will do well. For example the Kudzu vine has been successfully used in many parts of the United States. Given the right conditions, it puts out vigorous creeping runners which can completely cover the floor, sides and flanks of a gully with a dense blanket of vegetation. However its performance has been most disappointing in several countries in Africa (Hudson, 1971). A useful practice is to make trial plantings. Hudson (1971) and Bennett (1955) list some plants which have been used in various parts of the world and in America, respectively.

3.11 Grassed waterways

Grassed waterways are an economical and widely used means of conveying surface runoff from farm fields at non-erosive velocities. They provide protection against soil erosion and the vegetation in them (usually grasses or grass-legume mixtures) can be grazed or mown for hay. There are also a number of other hydraulic engineering situations in which conventional treatments for the stabilisation of surfaces subject to erosion by the intermittent flow of flood water are uneconomic or unaesthetic. Examples of these situations are dam emergency spillways and river flood banks.

3.11.1 General considerations

When considering whether grass can provide reliable long-term protection to earth slopes it must be borne in mind that if it is to perform satisfactorily, there must be time for it to recover between periods of immersion. Hence its suitability for intermittent flow situations.

Beside determining the probable degree of intermittency, the designer is concerned with the resistance to erosion, i.e. the length of time for which the grass surface remains stable, and also the hydraulic frictional characteristics which are pertinent to the overall channel dimensions.

U.S. Soil Conservation Service (1954) provides a list of permissible velocities that can be tolerated by a uniform grass cover, the limiting velocity being dependent on the grass type, hydraulic slope, and the erosional resistance of the soil. Generally the data indicates that 'long-term' stability will be maintained provided the velocity is $\lesssim 1$ m s^{-1}. Whitehead examined the available information and suggested that a well-chosen grass cover can withstand the following velocities: $\leqslant 2$ m s^{-1} for prolonged periods (e.g. > 10 h); 3-4 m s^{-1} for periods of several hours; $\leqslant 5$ m s^{-1} for brief periods (say, < 2 h). The limiting velocity is reduced as the quality of the cover decreases and Whitehead (1976) has summarised the experimental data in a velocity-duration of inundation diagram.

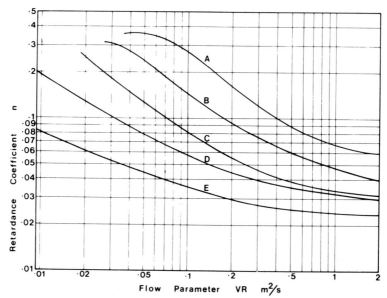

Fig. 3.19 Friction resistance of grasses; the letters by the curves refer to the retardance categories defined in table 3.10

The hydraulic roughness of a plant stand is a function of the physical characteristics of the grass, e.g. height, stiffness and density, and is very variable. Roughness is also a function of the flow depth and velocity. Despite this complexity it has been found that n, the Manning coefficient of roughness (known as the 'retardance coefficient' when applied to grassed channels) shows a distinct correlation with the product VR where V is the velocity and R the hydraulic radius (see section 2.6.1). This relationship is characteristic of the vegetation and practically independent of channel slope and shape (Chow, 1959). As a result the U.S. Soil Conservation Service (1954) developed a number of experimental curves for n versus VR (fig. 3.19) for five differing degrees of retardance; very high, high, moderate, low and very low as shown in table 3.10. These features permit design on the basis of the Manning formula (equation [2.51]).

Table 3.10 General guide to vegetal retardance categories

Stand*	Average length (cm)	Retardance category
Good	>76	A
	28−61	B
	15−25	C
	5−15	D
	<5	E
Fair	>76	B
	28−61	D
	15−25	D
	5−15	D
	<5	E

*Stand refers to the density of grass, e.g. number of stems per unit area

Once the particular type of grass for channel lining has been selected the hydraulic design is generally considered in two stages. The first stage is to design the channel for stability, paying particular attention to the peak discharge rate. Maximum velocities occur at the minimum retardance and this usually occurs when the grass height is least. Once a permitted design velocity has been selected, channel dimensions may be determined. The second stage is to consider the maximum capacity, i.e. to determine the depth of flow necessary to accommodate the design discharge at the maximum retardance; this usually occurs when the grass height is tallest. With these criteria, the general design procedure is as outlined in the following two sections.

3.11.2 Design for stability
With knowledge of the channel slope, discharge and grass character the following steps may be performed:
(a) Assume a value of n and determine VR for a given retardance class using fig. 3.19.
(b) Select the permissible velocity and compute the value of R.

(c) Manning's formula, equation [2.51], may be transformed to determine the value of the product $VR = R^{5/3} S^{1/2}/n$ and the resultant value may be checked against the value derived in (a).

(d) If the values of VR obtained from (a) and (c) do not match, select further trial values of n until correspondence is achieved; this represents a solution of Manning's equation which is consistent with fig. 3.19.

(e) Calculate the cross-sectional area of the water from $A = Q/V$.

(f) Since the correct values of A and R are known, the section dimensions may be determined (see example 3.2).

3.11.3 Design for maximum capacity

The next stage of the design is to determine the maximum depth consistent with a fully developed lining.

(a) Assume a depth z and calculate the water area A and hydraulic radius on the basis of the known channel shape and dimensions.

(b) Evaluate the velocity $V = Q/A$ and the product VR.

(c) Select the higher retardance class category (guided by table 3.10) and determine n using fig. 3.19 and the value VR from (b).

(d) Compute the velocity by equation [2.51] and check this value against the V obtained in (b).

(e) If the trial V from (b) fails to match the value deduced in (d), select a new trial value of the depth and repeat the calculation procedure until consistency is achieved.

(f) Add the recommended freeboard (additional depth from the top of the channel to the water surface to prevent overtopping by waves; see Chow, 1959, page 159).

3.11.4 Selection of grass

Once the designer has determined the channel dimensions, he then has to choose the grass which will provide a stable erosion-resistant surface. In making this decision it is also necessary to consider whether the grass area has a secondary purpose, e.g. grazing or recreation/amenity use, and also its subsequent management. For example, situations which do not call for agricultural or recreation/ amenity uses and where access is difficult, will benefit from slow-growing low-maintenance species. From the hydraulic point of view it is better to aim for grasses forming a uniform tightly-knit sod and to avoid tussock forming grasses which channel the flows and increase the chance of erosion. Beyond these, the selection depends mainly on the climate and soil in which the grass will grow and survive under the given conditions. Persistency is important and perennial grasses are recommended. No single grass species has all the characteristics which are desirable for a specific hydraulic engineering purpose. It is thus a common practice to utilise a mixture of species whereby the unique properties of each component can be exploited. For example a non-graminous nitrogen-fixing species (e.g. clover) might be sown to increase the nutritive value of the vegetation. One might also include rapidly establishing but poorly persisting species, e.g. Italian rye grass which will act as a nursecrop to protect the work in the early development of the grass sward and to provide cover for the slower developing long-term sward. In the U.K. it is common practice with water authorities to sow a grass mixture of an agricultural type which is based on perennial rye grass. Other minority species are added from the agricultural-feeding viewpoint. American experience has highlighted Bermuda grass as particularly suited to hydraulic protection works. It is also highly rated in other subtropical areas such as India and Australia. Where a low maintenance non-agricultural use is required, vigorous species such as perennial rye grass should be avoided and less vigorous species used, e.g. red fescue, hard fescue and brown top.

The guide by Whitehead (1976) provides excellent insight into the use of grass in hydraulic engineering, providing information about hydraulic features of grass linings together with design examples and practical details of the selection, establishment and management of grass mixtures.

Example 3.2 Design of grass-lined channel

Design a trapezoidal channel with side slopes 1:2 to carry 2.5 m³ s⁻¹ at 1:50 slope. The grass cover is expected to have a minimum retardance D and a maximum retardance B. Allowable velocity is chosen to be 1.5 m s⁻¹ at the lower retardance.

Following the procedures described in sections 3.11.2 and 3.11.3 a two-stage analysis is performed.

Design for stability
Taking the permissible velocity for design as 1.5 m s⁻¹ and using the n versus VR curve (fig. 3.19) the trial computations used in the design are as follows:

| Trial no. | n | VR $(m^2 s^{-1})$ | R (m) | $R^{5/3} S^{1/2}/n$ $(m^2 s^{-1})$ |
	(1)	(2)	(3)	(4)
1	0.04	0.28	0.19	0.22
2	0.03	1.85	1.23	6.66
3	0.035	0.56	0.37	0.77
4	0.038	0.36	0.24	0.34
5	0.037	0.40	0.27	0.42

Notes
Column (1) Trial values of roughness factor
Column (2) From fig. 3.19, Category D
Column (3) Hydraulic radius VR (column 2)/V (allowable velocity)
Column (4) Computed VR based on Manning's equation [2.51]
                with $S = 0.02$, n (from column 1) and R (from column
                3) (for comparison with column 2)

The correct values for the determination of the section are $R \doteq 0.25$ m
(interpolation of trial no.s. 4 and 5) and cross-sectional area $A =$
$Q/V = 2.5/1.5 = 1.67$ m^2.

For a trapezoidal channel of base width b and side slope 1:m (see
fig. 2.13)

$$A = bd + md^2 \tag{3.17}$$

$$P = b + 2\sqrt{(1 + m^2)}d \quad \text{(wetted perimeter)} \tag{3.18}$$

Since $R = A/P$ elimination of b from equations [3.17] and [3.18]
leads to the quadratic expression:

$$[2\sqrt{(m^2 + 1)} - m]d^2 - (A/R)d + A = 0 \tag{3.19}$$

Substituting $m = 2$, $A = 1.67$, $R = 0.25$ yields the solution $d = 0.28$
m. (*Note*: the other solution is rejected because it corresponds to
unobtainable conditions). From equation [3.17] $b = 5.4(3)$ m.

Design for maximum capacity
From the known data the trial calculations are given below.

Trial no.	z (m) (1)	A (m^2) (2)	P (m) (3)	R (m) (4)	V (m s^{-1}) (5)	VR (m^2 s^{-1}) (6)	n (7)	$R^{2/3} S^{1/2}/n$ (m s^{-1}) (8)
1	0.40	2.49	7.22	0.34	1.00	0.34	0.072	0.95
2	0.50	3.22	7.67	0.42	0.78	0.33	0.073	1.08
3	0.41	2.56	7.26	0.35	0.98	0.34	0.072	0.97

Notes

Column (1) Trial values of depth

Column (2) From equation [3.17] with $d = z$ and $b = 5.4(3)$ m

Column (3) From equation [3.18] with $d = z$ and $b = 5.4(3)$ m

Column (4) From A (column 2)$/P$ (column 3)

Column (5) From $V = Q/A$ (column 2)

Column (6) From V (column 5) $\times R$ (column 6)

Column (7) From fig. 3.19 using VR (from column 6)

Column (8) From Manning's equation [2.51] using R (column 4) and n (column 7) (for comparison with column 5)

The water depth at maximum capacity is 0.41 m. Adding a free board of 0.07 m, the total depth is 0.48 m and the top width is 7.3 m.

3.12 Vegetation in coastal protection

3.12.1 *Introduction*

Coastal protection — a guard against the ravages of the sea — has a long history. Early records show that the Frisians (or Coastal Dutch) built earth-mounds at least two thousand years ago to avoid loss of land by flooding. Dyking started about the AD 1000 and is now one of the most widely used methods of protecting low coastlines. Sea banks themselves require protection and this is usually achieved by a combination of grassing and rock revetments (on the seaward side). In some locations additional protection is gained from the presence of marsh banks which dissipate wave energy. Coastal dunes are the natural equivalent of sea banks, forming barriers between the sea and land. They occur in places with an active sand supply, strong onshore winds, low beach angles, large tidal ranges and low precipitation as important contributory conditions. Plants play an important role in dune formation and stabilisation. Sand-dunes by themselves have little inherent strength to resist storm attacks and must be wide enough to be partially expendable when under serious attack. To maintain dunes of sufficient width and to rebuild them after attack, it is essential to maintain an adequate cover of vegetation. Sand from which vegetation has been stripped can be rapidly shifted, piling up and threatening to bury adjacent land. Salt-marsh plantations also make a valuable contribution to the stabilisation of eroding shore-lines in many sheltered coastal areas. Where marsh banks have been lost, there is ample evidence, e.g. Newcombe *et al.* (1979), to indicate an acceleration of erosion, and sometimes the necessity of introducing artificial protection measures (see Whitehead, 1976).

Sand-dunes and salt-marshes possess many similar ecological features. In each situation, a particle-laden flow encounters a stand of vegetation within which the fluid velocity is markedly reduced (see section 3.2.1), providing the necessary conditions for the flow to shed its sediment load. The ensuing accretion changes the character of the environment paving the way for new species. In both habitats, the spatial extent of plant species is in a state of flux which with development gives rise to plant successions (see section 1.7); this is a major characteristic of both ecosystems and one which must be taken into consideration when planning biotechnical works. Whereas in normal soils there is usually a rich and diverse flora, in sand-dunes and salt-marshes the range is comparatively narrow and confined to well-adapted species.

To properly exploit the potential of plants in biotechnical works it is vital to aim for ecological stability. A first step is to define the environment in which a plant can or cannot thrive. As an example we shall consider the planting strategy within coastal dune systems. This theme was examined in Bache and MacAskill (1981) who also surveyed the interrelationship between biotechnical works and the environment of salt-marshes and sea banks.

3.12.2 Sand-dunes
Basic features of the zones occurring on sand-dunes are shown schematically in fig. 3.20. Here it should be stressed that the location, extent and nature of these zones varies widely.

The pioneer zone is the area of recent or continuing sand move-ment. The early stage of dune formation is known as the embryo dune phase and this occurs above the uppermost drift line where dunes are liable to submergence by peak spring tides. In north-west Europe, they are frequented by the perennial grasses, e.g. lyme grass and sand couch, these being able to withstand occasional wetting by seawater and readily survive in a saline environment.

By trapping sand, the dune level is raised and allows the invasion of the 'dune builders' of which marram grass and American beach-grass are well-known examples. Marram grass, as with many of the dune builders, is unable to withstand the salinity levels endured by the first colonisers and thrives best in an accreting situation. Ranwell (1958) states that marram grass can withstand accretion rates of up to 1 m per year over short periods; this contrasts with the species in the pioneer zone which are effectively limited to accretion rates of less than 0.3 m per year.

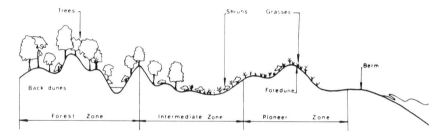

Fig. 3.20 Schematic representation of vegetation zones on a low sand-dune complex

While there are still areas of bare sand the dunes are said to be in the yellow phase (Chapman, 1976) and with a complete cover the dune is effectively stabilised. When the sand supply diminishes, e.g. due to the increased height of the vegetated dune surface, the dune passes into a mature or 'grey' dune phase, akin to the intermediate zone (fig. 3.20) and is normally considered as a progression in succession to the stable climax – forest. It is a highly variable and ill-defined area usually behind the active pioneer zone; it receives little sand supply and nutrient levels are low. The associated vegetation is extremely varied, including mosses, grasses and scrub (Chapman, 1976).

During the grey dune phase and its passing into forest, calcium supplies are washed out and the soil becomes progressively more acid. Forests form on dunes only after a substantial period of soil development and only on sites with considerable protection from salt spray and flooding. The type of woodland is varied and discussed in Chapman (1949).

Technical details of dune stabilisation work and methods are given in a number of comprehensive publications, e.g. Woodhouse (1978), Adriani and Terwindt (1974) and DIN 19 657. These, and others are supported by a wealth of experience with Chapman (1949) and van de Burgt and van Bendegom (1949) providing valuable discussion of the basic principles.

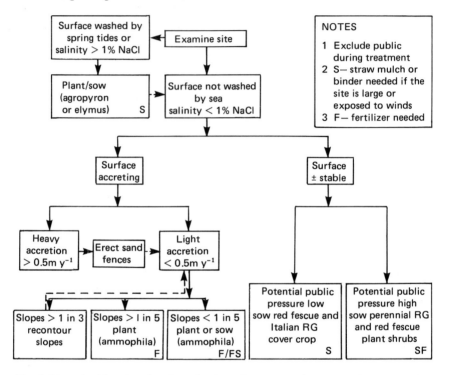

Fig. 3.21 A decision key for the selection of dune stabilisation techniques; solid arrows indicate primary choice of option; broken arrows indicate subsequent options; 'accretion' refers to actual sand input, taking account of erosion

As an aid for choosing the stabilisation technique and identifying the needs of a particular situation, Boorman (1977) produced the key shown in fig. 3.21 but he stressed that there is really no substitute for an adequate understanding of the basic concepts of dune processes.

DUNE DEVELOPMENT

Dunes can be developed following the principles which have been outlined making use of vegetation or sand fences to enhance deposition. Opinions vary on the location of a planned dune system. Generally the dune should be at a sufficient distance from mean high-water to avoid persistent wave attack. If this distance grows too small, the beach in front will not supply sufficient drift sand to replace the erosion caused by storm damage. If the distance is too large there is the risk that the quantity of drifting sand becomes too great and the dune grasses become overwhelmed. Knutson (1978) comments that foredune restoration is most likely to succeed when the new dune is formed along the natural vegetation line and associated aspects are discussed in Woodhouse (1978) and Adriani and Terwindt (1974).

To resist storm wave attacks and to provide nourishment to beaches, the dune complex must be sufficiently high and wide. In the United Kingdom a minimum width of 50 m is recommended where dunes are essential to coastal defence (Water Space Amenity Commission, 1980). The required height of dunes depends on a combination of tidal elevation and wave run-up, commensurate with an acceptable level of flood risk.

Fences are often used for trapping sand and are useful in that they can be placed at any time and are immediately operative. In areas subject to flooding they can be used to raise the elevation of an area prior to planting. Where sand is too mobile for vegetation to become established, fences form an effective windbreak and if a heavy accretion is expected fences must be used to avoid suffocation of the plants. For marram grass, Boorman (1977) selects the critical rate as 0.5 m y^{-1} (see fig. 3.21).

DUNE STABILISATION

Dune stabilisation as opposed to dune building is required where dunes are constructed mechanically or where vegetation has suffered extensive damage.

The principles and practices involved in planting grasses to build dunes generally apply to stabilising bare and unstable dunes, but there are differences in the plant spacing and planting pattern (Woodhouse, 1978). The sand surface can also be protected by netting, brush or stone, which together with mulches and chemical binders represent a number of options during the establishment of vegetation. Woodhouse (1978) reviews these methods, but stresses that vegetation is the only long-term stabiliser.

Where dunes have been extensively damaged by wind or water during severe storms, the corrective procedures depend on the type of damage (Knutson, 1978). For blowouts the entire disturbed area should be planted. For closing 'washover channels' caused by breaching during storms, dune levels should be restored using fences and then planted. The same point is emphasised by Chapman (1949) in his 'principles of dune stabilisation'. If reshaping of blown sand is to be carried out, it is important to grade the surface as indicated in fig. 3.21. The maximum slope should be set at 1:2 or less for planting, or 1:5 or less for mechanical sowing of marram grass (Adriani and Terwindt, 1974).

Differences of approach arise when stabilising outer dunes and inner dunes. For outer dunes, when the sand supply is normal, the dune-binding plants grow abundantly and can be generally used. For the inner dunes it is desirable to retain a complete plant cover to hold the surface. However, longer-term objectives should also be considered. The first is to prepare for the climatic climax, which can be selected with reference to the climate of the region and the neighbouring climatic climax vegetation, e.g. forest. The second concerns the intended use of the dune area.

In an accreting situation, grasses such as marram grass and lyme grass can continue to be used, although it should be noted that rabbits have an appetite for lyme grass and it should not be used on its own (Boorman, 1977). Where there is insufficient wind-borne sand to keep the dune grasses nourished, re-seeding with turf grasses is likely to be more effective. Details of the procedures for establishing dune pastures are given by the Countryside Commission for Scotland and are briefly described in Boorman (1977).

Chapman (1949) advocates the use of annual lupins, e.g. the blue lupin, as these increase the cohesion in the upper layers of sand by increasing the humus content. They will also act as a nurse to grasses and trees. Chapman (1949) also recommends that afforestation should be the final goal in nearly all dune reclamation work since a close covering of trees gives the greatest stability to an area. Trees cannot be planted until the dune has been stabilised because trees are not sand-binders (see Chapman, 1949), and Miers (1979) quotes situations which illustrate this view. Both Chapman (1949) and van de Burgt and van Bendegom (1949) review the criteria for selecting the appropriate type of trees, these covering: the climatic climax, the rapidity of growth, the tolerance to sea spray and soil moisture.

3.13 References

Adriani, M.J. and Terwindt, J.H.J. (1974) *Sand Stabilization and Dune Building*. Rijkswaterstaat Communications No. 19. The Hague, Netherlands.
Allen, S.E., Grimshaw, M., Parkinson, J.A. and Quarmby, C. (1974) *Chemical Analysis of Ecological Materials*. Oxford: Blackwell.
American Society of Civil Engineers (1975) *Sediment Engineering*. Manual No. 24. New York: American Society of Civil Engineers.

Bache, D.H. and MacAskill (1981) 'Vegetation in coastal and streambank protection'. *Landscape Planning* **8**, 363-85.

Bache, D.H. and Unsworth, M.H. (1977) 'Some aerodynamic features of a cotton canopy'. *Quarterly Journal of the Royal Meteorological Society* **103**, 121-34.

Bennett, H.H. (1955) *Elements of Soil Conservation*. New York: McGraw-Hill.

Binder, W. (1979) *Schriftenreihe Bayerisches Landesampt für Wasserwirtschaft.* Heft 10. Munchen: Grundzüge der Gewasserpflege.

Bonham, A.J. (1980) *Bank Protection Using Emergent Plants Against Boat Wash in Rivers and Canals*. Report No. IT 206. Wallingford, U.K.: Hydraulics Research Station.

Boorman, L.A. (1977) 'Sand-dunes'. In Barnes, R.S.K. *ed. The Coastline.* London: John Wiley.

Bradshaw, A.D. and Chadwick, M.J. (1980) *The Restoration of Land*. Oxford: Blackwell.

British Standard BS 6031. *1982 Code of Practice for Earthworks*. London: British Standards Institution.

Bruun, P. (1972) 'The history and philosophy of coastal protection'. *Proceedings of the Thirteenth Coastal Engineering Conference, Vancouver B.C., Canada.* Volume 1. New York: American Society of Civil Engineers, 33-74.

Burroughs, E.R. and Thomas, B.R. (1977) 'Declining root strength in Douglas fir after felling as a factor in slope stability'. U.S. Department of Agriculture Forest Service, Research Paper INT-190. Ogden, Utah: Intermountain Forest and Range Experiment Station.

Chapman, V.J. (1949) 'The stabilisation of sand-dunes' *Proceedings of the Conference on Biology in Civil Engineering*. London: Institution of Civil Engineers, 142-57.

Chapman, V.J. (1976) *Coastal Vegetation*. 2nd edn. Oxford: Pergamon.

Charlton, F.G. (1978) *The Hydraulic Geometry of Some Gravel Rivers in Britain*. Report No. IT 180. Wallingford, U.K.: Hydraulics Research Station.

Charlton, F.G. (1980) *River Stabilisation and Training in Gravel Rivers*. Report No. IT 217. Wallingford, U.K.: Hydraulics Research Station.

Charutamra, D. (1981) 'Vegetation and the stability of earth slopes'. M.Sc. thesis. Glasgow, Scotland: University of Strathclyde.

Chepil, W.S. (1960) 'How to determine required width of field strips to control wind erosion'. *Journal of Soil and Water Conservation* **15**, 72-75.

Chepil, W.S. and Woodruff, N.P. (1963) 'The physics of wind erosion and its control'. *Advances in Agronomy* **15**, 211-302.

Chow, V.T. (1959) *Open Channel Flows*. New York: McGraw-Hill.

Clouston, B. (ed.) (1977) *Landscape Design with Plants*. London: Heinemann.

Countryside Commission for Scotland. Reseeding of dune pastures. Information Sheet No. 5.2.2. Perth: Countryside Commission.

Dallaire, G. (1976) 'Controlling erosion and sedimentation on construction sites'. *Civil Engineering-ASCE* **47** October, 73-77.

Department of the Army (1978) 'Planting guidelines for marsh development and bank stabilisation'. *Engineer Manual*. No. 1110-2-5002. Washington, D.C.: Department of the Army, Corps of Engineers, Office of the Chief of Engineers.

DIN 18 918. *Landschaftsbau: Sicherungsbauweisen. Sicherungen durch Ansaaten, Bauweisen mit lebenden und nicht lebenden stoffen und Bauteilen, kombinierte Bauweisen*. Ausschuss, Berlin: Deutcher Normen.

DIN 19 657. *Sicherungen an Gewässern, Deichen und Küstendünen*. [Protection of water banks, dykes and coastal dunes]. Ausschuss, Berlin: Deutcher Normen. English translation available from British Standards Institution, London.

Donovan, R.P., Felder, R.M. and Rogers, H.H. (1976) *Vegetation Stabilization of Mineral Waste Heaps*. Report No. EPA-600/2-76-087. Washington, D.C.: U.S. Environmental Protection Agency.

Downing, M.F. (1978) 'Drainage'. In Hackett, B. *ed. Landscape Reclamation Practice*. London: IPC Science and Technology Press, 70-84.

Dunne, T. and Leopold, L.B. (1978) *Water in Environmental Planning*. San Francisco: Freeman.

Endo, T. and Tsuruta, T. (1968) 'On the effect of trees' roots upon the shearing strength of soil'. Annual Report of the Hokkaido Branch, Forest Experiment Station, Sapporo, Japan, 167-82. English summary.

Fleming, G. (1969) 'Design curves for suspended solid load estimation'. *Proceedings of the Institution of Civil Engineers* **43**, 1-9.

Forestry Agency, Japan (1976) 'Managing forests for water supplies and resource conservation'. In *Hydrological Techniques for Upstream Conservation*. F.A.O. Conservation Guide No. 2. Rome: Food and Agriculture Organisation of the United Nations, 1-12.

Gemmell, R.P. (1977) 'Reclamation and planting of spoiled land'. In Clouston, B. *ed. Landscape Design With Plants*. London: Heinemann, 179-208.

Gill, C.J. (1970) 'The flooding tolerance of woody species – a review'. *Forestry Abstracts* **31**, 671-.

Gill, C.J. (1977) 'Some aspects of the design and management of reservoir margins for multiple use'. *Applied Biology* **2**, 129-82.

Gill, C.J. and Bradshaw, A.D. (1971) 'The landscaping of reservoir margins'. *Landscape Design* **95**, 31-34.

Gray, D.H. (1970) 'Effects of forest clear-cutting on the stability of natural slopes'. *Bulletin of the Association of Engineering Geologists* **7**, 45-66.

Gray, D.H. (1978) 'Role of woody vegetation in reinforcing soils and stabilizing slopes'. *Proceedings of the Symposium on Soil Reinforcing and Stabilizing Techniques in Engineering Practice*. Sydney, Australia: The New South Wales Institute of Technology, 253-305.

Gray, D.H. and Leiser, A.T. (1982) *Biotechnical Slope Protection and Erosion Control*. New York: Van Nostrand Reinhold.

Gregory, K.J. and Walling, D.E. (1973) *Drainage Basin Form and Process*. London: Arnold.

Hackett, B. (1972) *Landscape Development of Steep Slopes*. Stockfield, U.K.:

Oriel.

Hagen, L.G., Skidmore, E.L. and Dickerson, J.D. (1972) 'Designing narrow strip barrier systems to control wind erosion'. *Journal of Soil and Water Conservation* **27**, 269-72.

Haslam, S.M. (1978) *River Plants*. Cambridge: Cambridge University Press.

Heede, B.H. (1977) 'Gully control structures and systems'. In *Guidelines for Watershed Management*. F.A.O. Conservation Guide No. 1. Rome: Food and Agriculture Organisation of the United Nations.

Holt, R.F., Timmons, D.R. and Burwell, R.E. (1979) 'Water quality obtainable under conservation practices'. *Universal Soil Loss Equation: Past, Present, and Future*. SSSA Special Publication No. 8. Madison, Wisconsin: Soil Science Society of America, 45-53.

Hudson, N. (1971) *Soil Conservation*. London: Batsford.

Hutnik, R.J. and Davis, G. eds. (1973) *Ecology and Reclamation of Devastated Land*. (2 volumes). London: Gordon and Breach.

Institution of Water Engineers (1969) *Manual of British Water Engineering Practice*. Volume II (4th edn.). Cambridge: Heffer.

Keown, M.P., Oswalt, N.R., Perry, E.B. and Dardeau, E.A. (1977) *Literature Survey and Preliminary Evaluation of Streambank Protection Methods*. Technical Report No. H 77.9. Vicksburg: U.S. Waterways Experiment Station.

Kite, D.J. (1980) 'Watercourses – open drains or sylvan streams'. *Trees at Risk*. Tree Council Annual Conference (in association with World Wildlife Fund) March 1980, London. Available from London: Tree Council.

Knutson, P.L. (1978) 'Planting guidelines for dune creation and stabilization'. *Proceedings of the Symposium on Technical, Environmental, Socioeconomic and Regulatory Aspects of Coastal Zone Management*. San Francisco, California. New York: American Society of Civil Engineers, 762-79.

Kohnke, H. (1950) 'The reclamation of coal mine spoils'. *Advances in Agronomy* **2**, 317-49.

Little, M.G. and Jones, H.R. (1979) *The Uses of Herbaceous Vegetation in the Drawdown Zone of Reservoir Margins*. Technical Report No. TR 105. Medmenham, U.K.: Water Research Centre.

Meyer, Y.D. and Kramer, L.A. (1969) 'Relation between land slope shape and soil erosion'. *Agricultural Engineering* **50**, 522-23.

Miers, R.H. (1974) 'The civil engineer and field drainage'. *Journal of the Institution of Water Engineers* **28**, 211-27.

Miers, R.H. (1977) *The Role of Vegetation in Land Drainage*. Report of a study tour undertaken in West Germany. London: Agricultural Development Advisory Service, Ministry of Agriculture Fisheries and Food.

Miers, R.H. (1979) 'Land drainage – its problems and solutions'. *Journal of the Institution of Water Engineers and Scientists* **33**, 547-79.

Mifkovic, C.S. and Petersen, M.S. (1975) 'Environmental aspects – Sacramento bank protection'. *Proceedings of the American Society of Civil Engineers* **101** (HY5), 543-55.

Miller, C.R. and Borland, W.M. (1963) 'Stabilization of Fivemile and Muddy creeks'. *Proceedings of the American Society of Civil Engineers* **89** (HY1), 67-97.

Ministry of Agriculture, Fisheries and Food (1979) *Fertiliser Recommendations*. Booklet No. GF1. London: Her Majesty's Stationery Office.

Newcombe, C.L., Morris, J.H., Knutson, P.L. and Gorbics, C.S. (1979) *Bank Erosion Control with Vegetation, San Francisco Bay, California*. Report No. MR 79-2. Fort Belvoir, Virginia: U.S. Army, Corps of Engineers, Coastal Engineering Research Center.

Parsons, D.A. (1963) 'Vegetative control of streambank erosion'. *Proceedings of the Federal Interagency Sedimentation Conference*. Miscellaneous Publication No. 970, Paper BEE No. 20. Jackson, Mississippi, U.S.: Agriculture Research Service, 130-36.

Pfeffer, H. (1978) 'Lebendbau und Landschaftspflege am Beispeil Untere Vils'. [Natural construction and care of the landscape exemplified by Untere Vils]. *Garten u Landschaft* **1/78**, 31-36.

Ranwell, D.S. (1958) 'Movement of vegetated sand-dunes at Newborough, Anglesey'. *Journal of Ecology* **46**, 83-100.

Rice, R.M. (1977) 'Forest management to minimize landslide risk'. *Guidelines for Watershed Management*. F.A.O. Conservation Guide No. 1. Rome: Food and Agriculture Organisation of the United Nations, 271-87.

Roehl, J.W. (1962) 'Sediment source areas, delivery ratios and influencing morphological factors'. *International Association of Scientific Hydrology*. Publication 59, 202-31.

Schiechtl, H. (1980) *Bioengineering for Land Reclamation and Conservation*. Alberta, Canada: The University of Alberta Press.

Schwab, G.O., Frevert, R.K., Edminster, T.W. and Barnes, K.K. (1966) *Soil and Water Conservation Engineering*. 2nd edn. New York: John Wiley.

Seginer, I. (1974) 'Aerodynamic roughness of vegetated surfaces'. *Boundary-layer Meteorology* **5**, 383-93.

Seibert, P. (1968) 'Importance of natural vegetation for the protection of the banks of streams, rivers and canals'. *Freshwater*. Nature and Environment Series No. 2. Strasbourg, France: Council of Europe, 33-67.

Sheng, T.C. (1977) 'Protection of cultivated zones: terracing steep slopes in humid regions'. *Guidelines for Watershed Management*. F.A.O. Conservation Guide No. 1. Rome: Food and Agriculture Organisation of the United Nations, 147-79.

Skidmore, E.L. and Woodruff, N.P. (1968) *Wind Erosion Forces in the United States and Their Use in Predicting Soil Loss*. Agriculture Handbook No. 346. Washington, D.C.: Agriculture Research Service, U.S. Department of Agriculture.

Tansley, A.G. (1968) *Britain's Green Mantle*. 2nd edn. Revised by Proctor, M.C.F. London: Allen and Unwin.

United Nations (1977) *Desertification: Its Causes and Consequences*. Oxford: Pergamon.

U.S. Department of Agriculture (1975) *Engineer Field Manual*. Washington, D.C.: Soil Conservation Service, U.S. Department of Agriculture, chapter 16.

U.S. Environmental Protection Agency (1975) *Methods of Quickly Vegetating Soils of Low Productivity, Construction Activities*. Report No. EPA-440/9-75-006. Washington, D.C.: Office of Water Planning and Standards, U.S. Environmental Protection Agency.

U.S. Soil Conservation Service (1954) *Handbook of Channel Design for Soil and Water Conservation*. Publication No. SCS-TP61. Washington, D.C.: U.S. Department of Agriculture.

U.S. Soil Conservation Service (1973) *Advisory Soils-6: Soil Erodibility and Soil Loss Tolerance Factors in the Universal Soil Loss Equation*. Washington, D.C.: U.S. Department of Agriculture.

U.S. Soil Conservation Service (1975) *Procedure for Computing Sheet and Rill Erosion on Project Areas*. Technical Release No. 51. Washington, D.C.: U.S. Department of Agriculture.

Van de Burght, J.H. and van Bendegom, L. (1949) 'The use of vegetation to stabilize sand dunes'. *Proceedings of the Conference on Biology and Civil Engineering*. London: Institution of Civil Engineers, 158-70.

Waldron, L.J. (1977) 'Shear resistance of root-permeated homogeneous and stratified soil'. *Journal of the Soil Science Society of America* **41**, 843-49.

Water Space Amenity Commission (1980) *Conservation and Land Drainage Guidelines*. London: Water Space Amenity Commission.

Whitehead, E. (1976) *A Guide to the Use of Grass in Hydraulic Engineering Practice*. Technical Note No. 71. London: Construction Industry Research and Information Association.

Wischmeier, W.H., Johnson, C.B., and Cross, B.V. (1971) 'A soil erodibility nomograph for farmland and construction sites'. *Journal of Soil and Water Conservation* **26**, 189-92.

Wischmeier, W.H. and Meyer, L.D. (1973) 'Soil erodibility in construction areas'. In *Soil Erosion: Causes, Mechanisms, Prevention and Control*. Special Report 135. Washington, D.C.: Highway Research Board, 20-29.

Wischmeier, W.H. and Smith, D.D. (1962) *Soil Loss Estimation as a Tool in Water Management Planning*. International Association of Scientific Hydrology, Publication No. 59, 148-59.

Wischmeier, W.H. and Smith, D.D. (1978) *Predicting Rainfall Erosion Losses – A Guide to Conservation Planning*. Agriculture Handbook No. 537. Washington, D.C.: U.S. Department of Agriculture.

Wolman, M.G. (1967) 'A cycle of sedimentation and erosion in urban river channels'. *Geografiska Annaler* **49**, 385-95.

Woodhouse, W.W. (1978) *Dune Building and Stabilization with Vegetation*. Report No. SR-3. Fort Belvoir, Virginia: U.S. Army, Corps of Engineers, Coastal Engineering Research Center.

Woodruff, N.P. and Siddoway, F.H. (1965) 'A wind erosion equation', *Soil Science Society of America, Proceedings* **29**, 602-08.

Wu, T.H., McKinnell, W.P. and Swanston, D.N. (1979) 'Strength of tree roots and landslides on Prince of Wales Island, Alaska'. *Canadian Geotechnical Journal* **16**, 19-33.

4 Plants and land disposal of wastes

4.1 Introduction

Land is the recipient of a major portion of the wastes we produce, disposal taking the form of landfill, spoil tips or by surface application; in this chapter we examine the latter. Emphasis is placed on the interaction between waste constituents and the soil-plant system in order to determine and sustain disposal rates; it also illustrates the interaction between plants and their chemical environment. Particular attention is paid to liquid and semi-solid wastes, of which sewage effluent and sewage sludges are important examples. Other examples include industrial wastes, leachates from landfill sites and agricultural wastes. The principles discussed also apply to irrigation waters.

Before the advent of the present-day sewage treatment plants – relying on rivers and coastal waters for ultimate discharge – sewage farms were widely used in both Europe and the United States, but gradually fell into disfavour. In recent years, however, land application techniques have attracted considerable attention because of the potential savings offered in terms of construction costs and energy requirements when compared to advanced waste-water treatment (Pound *et al.* 1978). Formerly, land application was considered mainly as a means of waste-water disposal. Today, however, there is a greater emphasis on land application being regarded as a waste-water treatment system – leading to renovation and re-use (Uiga *et al.* 1976). In the United States, the passage of the Federal Water Pollution Act Amendments in 1972 (P.L. 92-500) has had a significant impact on the acceptance of the land disposal of waste-waters. Section 201 of that legislation requires the study of alternative waste management techniques including land application, for the best practical treatment technology; it encourages recycling of waste-water and treatment system management that combines open space and recreational considerations.

Land treatment is sometimes thought of as being suitable for only food and domestic wastes. However an environmentally acceptable rate of application to a plant-soil system can be determined *for any and all industrial waste constituents*, with the possible exception of radioactive species (Overcash and Pal, 1979). In this respect, land treatment offers an innovative and possibly a future approach to industrial waste management.

Aside from economic factors (discussed in Loehr *et al.* 1979 and Culp *et al.* 1978) the major problems associated with land application arise from public acceptance, the contamination of ground and surface waters, pathogens, phytotoxicity due to overloading with heavy metals, and the contamination of the food chain by toxic elements.

Land treatment can also provide social and environmental benefits, such as the reclamation of sterile, disturbed and marginal soil, or the creation of green belts (Pound *et al.* 1978).

The success of a land application scheme depends on an intimate knowledge of the interaction of a constituent with the plant-soil system and where failures occur these are often attributable to the neglect of site data in the design. Of these the nature of the soil is one of the most important site-specific characteristics.

Experience has shown that successful land treatment sites are generally dependent on a thriving vegetative cover. A good vegetative cover improves the ability of soils in several ways; rooted plants encourage and maintain infiltration; transpiration promotes soil drying and aids the maintenance of aerobic conditions; under certain conditions the vegetative cover can be used to recover plant nutrients — when harvested it contributes to the removal of waste constituents and provides an economic return. The presence of plants provides an active ecosystem which aids the biological degradation of waste materials. The vegetative cover is also an effective indicator of the satisfactory waste assimilation by a soil-plant system and overall, the ability to maintain a viable crop or forest production is roughly equivalent to acceptable rates of waste application.

4.2 Perspective of land treatment systems

The principal design objective of a land application system is to determine the rate at which waste constituents — either as individuals or in combination — may be applied, and over what period this may be sustained. This is subject to the proviso that the land is not irreversibly damaged and leads to an environmentally acceptable assimilation of waste.

4.2.1 Plant-soil assimilative capacity

The assimilative capacity of land depends on contact and the processes which take place in the upper zone of the soil-plant system; these include adsorption, immobilisation, dispersion, chemical and biological decomposition. In spite of these widely differing processes, Overcash and Pal (1979) suggested that it is possible to view the assimilation capability as conforming to three patterns of behaviour described below.

DEGRADATION

For waste constituents that undergo decomposition or substantial crop uptake, the assimilative capacity is related directly to the rate of compound transformation or crop removal. Oils, grease and nitrogenous compounds are examples within this category.

Usually there is also an upper limit in the concentration of a constituent within the plant-soil system without incurring adverse plant responses, such as reductions in crop yield (see fig. 1.6). Once applied at or below the critical dosage level, the constituent degrades and no further dose should be applied until its concentration in the soil has decreased to about 10-20% above the background level. The resultant pattern of application and soil concentration is illustrated in fig. 4.1; this shows a saw-toothed pattern and depicts the design data required, i.e. critical dose levels and the rate of decomposition.

ACCUMULATION

This refers to compounds, e.g. heavy metals, which are relatively immobile and non-degradative, and are permitted to accumulate in soils until a critical concentration is reached. The critical level is usually associated with phytoxicity, or perhaps by constraints determined by entry into the food chain. Figure 4.2 illustrates that the lifetime of a particular site depends on the application rate and the type of metal.

MIGRATION

The last category refers to waste constituents which are non-degradative and migrate with water movement in the soil, e.g. deep percolation. The assimilation capacity is based on using sufficient land so that the concentration of the constituent in the recipient water, i.e. groundwater or surface water, conforms with water quality standards. Inorganic cations and anions fit into this category.

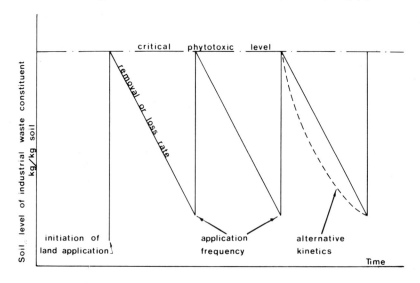

Fig. 4.1 Typical soil concentration response pattern for land application of organic constituents

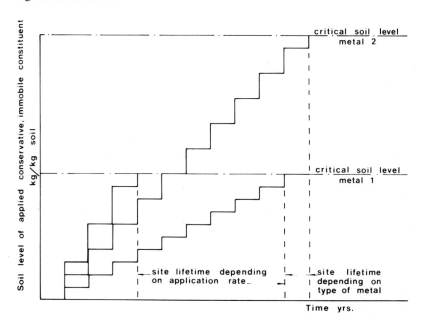

Fig. 4.2 Accumulation pattern and critical soil level for various metal application rates and metal types in industrial waste applied to land

4.2.2 Land limiting constituent analysis

With knowledge of the assimilative capacity for individual or combinations of constituents, a widely used approach is to determine the land area requirement commensurate with their rate of supply. This is determined by the ratio of supply rate (kg y^{-1}) to the assimilative capacity (kg ha^{-1} y^{-1}) as shown in table 4.1.

Table 4.1 Identification of the land limiting constituent

Waste steam constituent	Effluent (kg y^{-1})	Assimilative capacity (kg ha^{-1} y^{-1})	Area requirement based on site assimilative capacity and waste flow (ha)
H_2O	10^8	3×10^7	3.3
B.O.D.*	24 000	245 000	0.1
S.S.**	24 000	67 000	0.4
P	1 000	250	4.0
N	4 000	340	11.8†
Cu	8	9	0.9
Cd	0.8	0.17	4.7

*Biochemical Oxygen Demand (B.O.D.)
**Suspended Solids (S.S.)
†Land Limiting Constituent (L.L.C.)

The constituent requiring the largest land area is referred to as the land limiting constituent (L.L.C.). When a design is based on the L.L.C. of a waste, it automatically satisfies the land area requirements of the remaining constituents because these are smaller. The L.L.C. approach is very useful in that it rapidly identifies which constituents are critical to the design, perhaps deserving pre-treatment, and also for comparing the area requirements engendered by differing disposal sites (see Overcash et al. (1978) for a good case-book example). Although land area requirements can be reduced by pre-treatment (which reduces the supply rate), this is not usually possible for water.

4.2.3 Water character categories

To facilitate consideration of the many constituents present in water, Overcash and Pal (1979) split these into eight broad categories:

water or hydraulic loading;
phosphorus and sulphur compounds;

oil and grease constituents;
specific organic compounds;
acids, bases and salts;
anions;
metals;
nitrogen.

Within each category there are many compounds for which the plant-soil assimilation capacity is established; the broad categories indicate the general response to those compounds.

Individual categories are surveyed in Overcash and Pal (1979). In the present context, discussion is restricted to those most closely associated with the disposal of municipal waste-water, sewage sludges and agricultural wastes; they also illustrate the assimilation patterns shown in figs. 4.1 and 4.2.

4.3 Methods of land application

Methods of land application are varied and in the first instance depend on the solids content of the waste. Low-moisture materials, thickened slurries and sludges are generally dispersed by vehicle. For liquid wastes ($\lesssim 3\%$ solids) there are three major types of land treatment: slow rate (irrigation); overland flow; and rapid infiltration — the classification depending largely on the soil's permeability. The design of these systems is detailed in U.S. Environmental Protection Agency (1977) and shown schematically in fig. 4.3.

4.3.1 Slow infiltration

Slow infiltration is similar to crop irrigation in that they are each characterised by similar loadings and performance, but the objectives are different. In slow infiltration the aim is to maximise the application rate for a given effluent quality; the application rate depends on the hydraulic and renovative capacity of the soil. In crop irrigation, the object is to maximise crop yield and the application depends on the crop requirements.

Crop irrigation encompasses the irrigation of forests, landscape and turf areas as well as food and forage crops, whereas with slow infiltration, grasses are generally used because of their nitrogen uptake. Inspection of table 4.2 indicates that vegetation makes an important contribution to the removal of the nitrogen, phosphorus and the major cations.

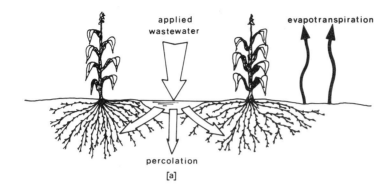

percolation
[a]

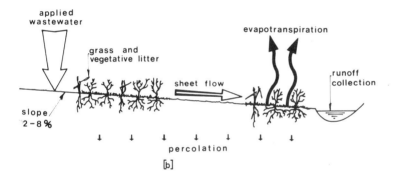

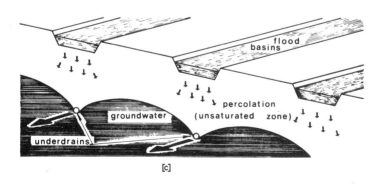

Fig. 4.3 Land treatment schemes showing the hydraulic pathway for (a) slow rate, (b) overland flow and (c) rapid infiltration systems

Table 4.2 Land disposal renovation mechanisms and their relative predominance in spray irrigation (S.I.), overland runoff (O.R.) and rapid infiltration (R.I.) systems*

Effluent constituent	Plant uptake**			Microbial degradation and/or immobilization†			Gaseous losses††			Adsorption precipitation			Ion exchange			Mechanical retention		
	S.I.	O.R.	R.I.	S.I.	O.R.	R.I.	S.I.	O.R.	R.I.	S.I.	O.R.	R.I.	S.I.	O.R.	R.I.	S.I.	O.R.	R.I.
B.O.D.	0	0	0	4	4	2	0	0	0	2	2	2	0	0	0	4	3	4
Suspended solids	0	0	0	4	4	2	0	0	0	2	2	2	0	0	0	4	3	4
Nitrogen	4	3	0	2	2	1	4	4	4	1	1	0	2	1	1	1	0	1
Phosphorus	4	3	0	1	1	0	0	0	0	4	4	4	1	1	1	1	0	1
Trace metals	1	1	0	2	2	1	0	0	0	4	4	4	1	1	1	0	0	0
Organic toxicants	0	0	0	3	2	1	1	1	1	4	4	4	0	0	0	0	0	0
Bacteria and viruses	0	0	0	3	4	3	0	0	0	2	2	2	0	0	0	4	3	4
Major cations	4	3	1	0	0	0	0	0	0	4	4	4	2	1	2	0	0	0
Major anions	2	3	0	0	0	0	0	0	0	4	4	4	0	0	0	0	0	0

*A 0-4 scale is used to express the relative importance of the proposed mechanisms and is derived from ranking the processes controlling the soil solution composition of a given component: 0 indicates that the process is inoperative or relatively insignificant, whereas 4 designates the major mechanism

**Removal by harvest crops

†Includes decomposition of organic material and loss of CO_2 to atmosphere and kill or other inactivation of pathogenic organisms

††Loss to atmosphere of: (a) N_2 and N oxides formed by biological and/or chemical denitrification; (b) volatile organic compounds; and (c) NH_3.

Slow infiltration systems are the most widely used of major types in land treatment, with many communities in the United States using this approach.

4.3.2 Overland flow

This is a biological treatment technique which consists of applying water at the top of a slope of relatively impermeable soil and allowing it to flow through vegetation to runoff collection ditches. It is used where soil permeability is low and other methods are not feasible. A well-known land treatment system incorporating overland flow has been in operation at Werribee Farm, treating waste-water from Melbourne, Australia since 1905 (see Anon, 1973).

Sod crops are well suited to overland flow systems because they allow very little erosion. Table 4.2 indicates that vegetation plays a significant role in the removal of nitrogen, phosphorus, the major anions and cations but it is generally less effective than in slow infiltration.

4.3.3 Rapid infiltration

In rapid infiltration land treatment most of the applied waste-water percolates through the soil and the treated effluent eventually reaches the groundwater. Very large amounts of waste-water (up to ~ 100 m applied per year) are either sprayed or flooded over an infiltration basin on highly permeable soils. Vegetation is not usually used, but there are exceptions.

4.4 Waste renovation processes

Waste treatment mechanisms may be broadly classified as physical, chemical and biological, but in many instances these are interdependent. Introductory accounts of these phenomena are given in Loehr et al. (1979), Overcash and Pal (1979) and Rich (1973). The relative significance of the renovation mechanisms in spray irrigation, overland flow and rapid infiltration is shown in table 4.2.

4.4.1 Physical processes

As a liquid waste enters and passes through the soil surface, suspended solids are removed by filtration. Particles with dimensions exceeding the soil pore sizes are removed by mechanical straining. Suspended material smaller than the pore sizes are also efficiently removed as a result of sedimentation, interception, inertial impaction, electrical forces and other trapping mechanisms reviewed in Ives and Gregory (1967).

The principal problem associated with suspended solids (S.S.) is their potential to cause surface clogging and reduce the hydraulic conductivity. Data from Laak (1970) indicates that for a variety of soils loadings of $\sim$ 225 kg S.S. ha^{-1} day^{-1} will not cause significant clogging. Sopper and Richenderfer (1979) reported findings in which the percolation capacity of agricultural and forest soils was substantially unaltered, whereas the surface infiltration capacity was affected.

4.4.2 Biological transformation

Biological processes resulting from a wide range of soil organisms play a dominant role in the decomposition of organic compounds and in the transformation of certain inorganic groups.

Organic substrates represent a class of matter through which enzymatic breakdown releases energy (termed respiration) for supporting metabolic activity and enables the organism to incorporate organic components into cellular material (termed synthesis). Respiration is an oxidation process and can be represented by the reaction:

$$C_aH_bO_cN_dP_eS_f \ + \ (O_2) \ \longrightarrow \ \begin{matrix} (CO_2) \\ (NO_3^-) \\ (H_2O) \\ (PO_4^{3-}) \\ (SO_4^{2-}) \\ (H^+) \end{matrix} \ + \ energy \quad [4.1]$$

organic substrate   (enzyme)

The rate at which organic matter degrades depends in part on the composition of the organic matter, but is critically dependent on the presence or absence of oxygen. In aerobic conditions, oxygen can be supplied as free molecular oxygen (O_2); whereas in anaerobic conditions, the organisms depend upon chemically bound oxygen. Anaerobic decomposition is generally slower than aerobic degradation.

In aerobic conditions a large fraction of the organic carbon (up to 60%) is released as carbon dioxide (CO_2). Much of the remaining carbon is incorporated into microbial cells.

In anaerobic conditions carbon dioxide and methane (CH_4) are released, but the bulk of the carbon remains in partially decomposed intermediates, such as organic acids and alcohols. Other gaseous end products include nitrogen (N_2) and nitrous oxide (N_2O) which result from the conversion of ammonium ions (NH_4^+) and nitrates (NO_3^-) together with hydrogen sulphide (H_2S) which is derived from sulphates (SO_4^{2-}).

MINERALISATION AND IMMOBILISATION
The process of converting organic matter into inorganic forms as illustrated in equation [4.1] is termed mineralisation and the resultant compounds, depending on their nature, are subject to gaseous loss, leaching, plant assimilation and to microbial and chemical conversion.

Microbial assimilation and transformations of the inorganic forms of nitrogen, phosphorus and sulphur to less available organic or biological forms is referred to as immobilisation.

4.4.3 Absorption of elements by plants
Mineral nutrients are usually taken up by plants in the ionic form and generally there are two mechanisms of absorption: First, nutrients may be carried into the plant as it takes up water from the soil solution. Second, during respiration, CO_2 is released by the plant roots. This combines with water to form carbonic acid (H_2CO_3) which dissociates to form hydrogen ions (H^+) and bicarbonate ions (HCO_3^-). When this happens, the H^+ in or at the root surface can exchange with cation nutrients which are in solution or adsorbed on the soil colloid (Sopher and Baird, 1978), and HCO_3^- can exchange with anion nutrients.

4.4.4 Chemical processes
The two chemical processes which account for most of the retention are precipitation and adsorption. Adsorption refers to the attachment of dissolved ions or compounds to the surface of the soil colloids, whereas precipitation results from the interaction of ions in solution to produce a new solid phase.

PRECIPITATION
When the concentration of anions and cations in solution becomes sufficiently high, insoluble or slowly soluble compounds may be formed due to the association of specific ions. In general the concentration of the dissolved constituents cannot exceed the concentrations determined by the solubility product K_{sp} of the corresponding compounds. For example, the solubility product of calcium carbonate ($CaCO_3$) is 4.82×10^{-9}. This means that if the product of

calcium concentration $[Ca^{2+}]$ and carbonate concentration $[CO_3{}^{2-}]$ exceeds 4.82×10^{-9}, $CaCO_3$ will precipitate out of solution. (*Note*: concentrations are expressed in moles/l). The magnitude of the solubility product reflects the solubility of compound. For example, the K_{sp} value for calcium sulphate $(CaSO_4)$ is 2.4×10^{-5} and this shows that calcium carbonate is less soluble than calcium sulphate.

The concentration of anions and cations in the soil solution can be altered in a variety of ways, thus affecting the opportunity of precipitation. Evaporation removes soil water thereby increasing concentrations, whereas the diluting effect of rainfall reduces concentrations. Adsorption or biological activity represent other mechanisms by which ionic concentrations may be altered.

ADSORPTION

Adsorption manifests as ion exchange and specific adsorption, depending on the bond strengths. In general, soil surfaces carry a net negative charge associated with clay and humic particles. This implies that the retention of some anions such as nitrate, chloride and sulphate is low and cationic exchange predominates.

CATIONIC EXCHANGE

Because of their surface negative charge, soil particles attract positively charged cations. Cation exchange takes place when one of the soil solution cations replaces one of the cations held on the surface of the soil colloid.

The ability of an ion in solution to displace dissimilar ions from the soil exchange surface depends on the relative concentration of the ions in solution and on its position in the sequence:

$$Al^{3+} > H^+ > Ca^{2+} > Mg^{2+} > K^+ > NH_4^+ > Na^+$$

which reflects the order of the 'intensity of attraction' between the ion and soil colloids (Loehr *et al.* 1979). The sequence indicates that when there are equivalent amounts of two ionic species, say calcium and sodium, distributed between the exchange sites and in solution, then the equilibrium state is such that calcium ions dominate the exchange sites due to their greater intensity of attraction (fig. 4.4). However, if there exists a disproportionately high concentration of one particular ion, say sodium, this then offsets the effect of the intensity of attraction and enables sodium ions to dominate the exchange sites (fig. 4.4(b)).

The sum total of exchangeable cations that a soil can adsorb is measured by the cation exchange capacity (C.E.C.) and is discussed further in section 1.5.

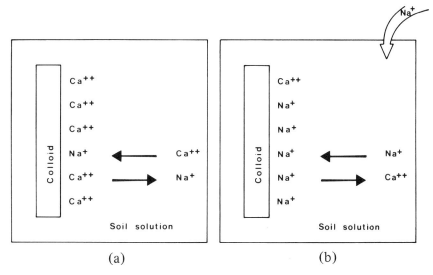

(a)   (b)

Fig. 4.4 (a) The intensity of attraction between soil colloids and specific cations influences the cation exchange process; Ca^{2+} is more strongly attracted and thus dominates the colloidal charge sites; (b) concentration can offset the effect of intensity of attraction; the greater concentration of Na^+ enables it to dominate the colloidal charge sites

SPECIFIC ADSORPTION

Soils possess the capacity to retain anions and cations so tightly that they are not exchangeable, a process termed specific adsorption; this mechanism dominates the retention of the heavy metal cations such as Zn, Cu, Ni, Cd, Pb, and some anions such as phosphate.

4.5 Water factors

The hydraulic capacity of a soil to receive and transmit water is an important component in the design and operation of an industrial treatment system. Without water movement nutrient transport into and through the soil profile would be impossible. Conversely, too rapid movement may lead to groundwater pollution.

The determination of the hydraulic assimilative capacity focuses on two factors: (a) the periodic loading; and (b) the application rate.

The maximum hydraulic loading can be assessed using the water balance equation (sections 2.1.2 and 2.7.1):

$$W + P = E + D + R \qquad\qquad [4.2]$$

where W is the waste-water load, R the runoff, D the vertical and lateral percolation through the soil, E the evapotranspiration and P the precipitation. For systems which depend on infiltration (as distinct from overland flow), runoff is zero, thus assuring minimal waste of water and good contact between the waste and the plant-soil system. The water budget is usually solved for W over monthly intervals and depends on knowledge of P, E and D.

Climatic data are required to determine the amount and frequency of precipitation events and the occurrence and extent of freezing conditions. Evapotranspiration is influenced by both climatic and plant features and is best calculated by the Penman method (section 2.3) or, if the necessary data are lacking, using the less sophisticated Blaney-Criddle approach (see U.S. Soil Conservation Service, 1970).

Percolation, the movement of the water through the soil, should be distinguished from infiltration, which refers to the movement of water through the soil surface (section 2.7). Percolation rates are difficult to assess and require *in situ* measurements. For a wide range of soil and vegetation conditions, experience suggests that the monthly percolation rate usually does not exceed 5-25% of the measured infiltration rate after prolonged wetting to saturated conditions using clean water (Overcash and Pal, 1979 and U.S. Environmental Protection Agency, 1977). Rough estimates of permeability (the rate of water movement through a unit cross-section in a saturated soil) may be obtained using table 2.10 with adjustments included for slope. Infiltration varies from one crop to another and depends on the surface condition. Tables 2.10 and 2.11 indicate that the vegetation cover with the best infiltration characteristics is undisturbed forest land.

From tables 4.1 and 2.10 it is evident that the soil-plant system offers considerable potential for assimilating water and is not usually land limiting except where rapid infiltration is required. In practice, waste-water should be applied at a rate designed to optimise the renovation capacity of the soil-plant system and to make best use of the nutrients within the waste.

Generally the soil-plant system responds better to periodic wetness than to continuous levels of wetness. The maximum productivity of plants is associated with maximum transpiration, and this can be approached by providing irrigation to balance the deficit between precipitation and actual evapotranspiration. In normal irrigation practice for crop production, water is applied to restore the soil moisture to field capacity when the available water is reduced to approximately 50%; this approach to replenishment is based on water shortage. However in application systems geared toward water disposal, the object is to maximise the flow of water. One approach (illustrated in fig. 4.5) is to ensure that evapotranspiration remains at its potential value by scheduling water delivery when approximately 20% of the available moisture is depleted (Loehr *et al*. 1979). Otherwise, if water is delivered at a rate greater than the balance between the evaporative demand and precipitation, the volume of percolating water will increase. Excessive water in the soil can inhibit plant growth (section 1.4) and lead to the leaching of waste constituents. Where the waste-water has a high nitrogen content it may be necessary to control the soil moisture to create conditions favouring denitrification (section 4.6).

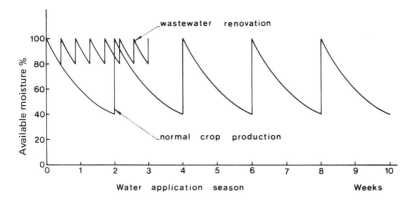

Fig. 4.5 Irrigation scheduling at shorter intervals will maintain a higher percentage of available moisture in the soil profile for evapotranspiration

Plant tolerance to soil moisture differs according to species and the stage of growth. For example, most crops are especially sensitive when they approach flowering. In general, grain crops such as wheat, oats and barley will suffer high losses in yield when subjected to soil saturation. Fodder and forage crops are generally more water tolerant, particularly the grasses. Amongst these, reed canary grass, tall fescue and timothy appear to be best adapted to moist conditions (Loehr *et al*. 1979). Trees and shrubs also differ in their tolerance to the level of the water table and flooding and further detail is given in van't Woudt and Hagan (1957).

4.6 Nitrogen

Nitrogen, an essential plant nutrient, is an important component of municipal and agricultural wastes and is often the factor limiting their application on land. Large inputs of nitrogen, while stimulating plant growth can lead to increased losses to ground and surface waters; this contributes to excessive growths of algae and weeds. Further, a build-up of nitrate in drinking water poses a health hazard − particularly to infants, and livestock. Ideally, the nitrogen application rate should be consistent with the nitrogen needs and utilisation rates of plants, so that the nitrogen excesses causing environmental damage are minimised.

4.6.1 Forms and processes
In order to maintain permissible levels of nitrogen in groundwater, management decisions on loading depend on: knowledge of the various forms of nitrogen in wastes; their properties; and the transformations which take place in the soil-plant system.

The forms of nitrogen of importance in waste management are outlined below:
(a) *Organic nitrogen* is bound in organic compounds such as proteins. Within the soil organic-N is not available for direct plant uptake, but is slowly transformed to available forms by means of microbial and chemical decomposition.
(b) *Ammoniacal nitrogen* includes two forms, the ammonium ion (NH_4^+) and gaseous ammonia NH_3. Ammonium ions, by virtue of their positive charge, are attracted to the negatively charged clay and organic colloids in the soil; this greatly restricts their movement by percolating waters. Plants and microorganisms make use of ammonium ions as a source of nitrogen. Ammonia gas exists in equilibrium with NH_4^+ and may escape into the atmosphere in significant quantities from soils that are neutral or alkaline.
(c) *Nitrite* (NO_2^-) is a highly mobile anion formed in soils as an intermediate during the conversion of ammonium ions to nitrate; it is toxic to higher plants in very small quantities, but is very rapidly oxidised to nitrate.
(d) *Nitrate* (NO_3^-) is a highly mobile ion, readily used by both plants and micro-organisms. Nitrate is of special interest because it is easily leached. The U.S. Environmental Protection Agency and World Health Organisation has recommended a drinking water standard of not more than 10 mg l^{-1} NO_3-N (nitrate nitrogen).

Other forms of nitrogen include nitrous oxide (N_2O) which is relatively unimportant and molecular nitrogen (N_2), which is a gaseous form comprising 80% of the normal atmosphere.

From the above, it is apparent that the nitrogen available to plants and for leaching is in the inorganic state. To discuss effectively the removal of nitrogen from waste material, it is important to understand something of the complex nitrogen transformations that may occur in the soil, since these contribute to permissible loading rates.

Ammonification

Ammonification is the conversion of organic-N into ammonium by soil microbes. The process is important for converting nitrogenous compounds into forms which are accessible to plants.

Nitrification

Nitrification is a two-stage process in which ammonium nitrogen (NH_4-N) is converted to nitrate nitrogen via the steps $NH_4^+ \rightarrow NO_2^- \rightarrow NO_3^-$. Both steps require oxygen and are thus favoured by well-aerated soils with a moisture content within or slightly below the acceptable range for higher plants (see section 1.4). The process essentially ceases below $0°C$ and is greatest for temperatures in the range $26-33°C$ (Brady, 1974) and in soils with a slightly acid to slightly alkaline pH. Nitrification is usually rapid compared with ammonification. Its major significance in waste application systems derives from the conversion of the relatively immobile form (NH_4^+) to the highly mobile form (NO_3^-) which is readily leached. Additionally, nitrate is required in the process of denitrification.

Mineralisation and immobilisation

Mineralisation is a general term covering the series of reactions involved in the transformation of organic-N into inorganic forms. Its most important processes are ammonification and nitrification.

Immobilisation is the process by which inorganic-N is converted into organic-N and is essentially the reverse of mineralisation.

The ratio of the carbon to nitrogen (C:N) in the decomposing organic matter indicates whether mineralisation or immobilisation is likely to take place. When materials with a high C:N ratio (low nitrogen concentration) are applied to soils immobilisation occurs, whereas mineralisation predominates when materials with a low C:N ratio (high nitrogen concentration) are added. Soil organic matter has a C:N ratio of about 10-12; the break-even point between mineralisation and immobilisation occurs at a C:N ratio in the range 20-32.

Nitrogen fixation

Nitrogen fixation refers to the change in inorganic-N into organic form, and is distinct from immobilisation in that its source is atmospheric nitrogen (N_2) rather than mineral or inorganic combined form. It is closely associated with the legumes and can represent an important input of N into the soil. However, it is generally only a minor input in land receiving waste application (Loehr *et al.* 1979).

NITROGEN LOSSES

Nitrogen is lost from the soil by several means including ammonia volatilisation, denitrification, crop uptake and harvesting, surface runoff and leaching.

Volatilisation

In moist alkaline soils, ammonium salts undergo a chemical reaction resulting in the evolution of ammonia gas. If an industrial waste is applied at the soil-plant surface, e.g. by surface irrigation, there is little to inhibit ammonia movement and a significant fraction of NH_4-N is lost. When wastes are incorporated into the soil by slow or direct injection, the loss is substantially reduced with the magnitude of the retention being proportional to the cation exchange capacity (Overcash and Pal, 1979).

Denitrification

This is the process by which nitrate-N and nitrite-N are converted into the gaseous forms N_2O and N_2 via the sequence $NO_3^- \rightarrow NO_2^- \rightarrow N_2O \rightarrow N_2$. These reactions take place in poorly aerated, waterlogged soils (i.e. essentially anaerobic conditions) or in anaerobic pockets and require the presence of available carbon from decomposing organic matter; denitrification is difficult to predict.

Leaching and runoff losses

On industrial waste land application sites efforts must be made to control the runoff and leaching losses of nitrogen. Runoff occurs whenever water application rates exceed the soil infiltration capacity (see sections 2.3 and 2.4) and are usually minimal in a well-designed and managed system. Leaching occurs when the water-holding capacity of the soil is reached and additional water percolates through the soil (see section 4.8) transporting the mobile constituents.

Crop uptake and harvest

Nitrogen uptake by crops followed by harvesting constitutes a significant mechanism for nitrogen removal. Without harvesting, the annual
nitrogen uptake will simply return to the soil in organic form. Some
representative values are shown in table 4.3. Grasses, particularly the
perennials, tend to have higher uptake rates than field crops. Although
considerable amounts of nitrogen may be taken up by trees during
the growing season, much of it is deposited annually in leaf and
needle litter.

Table 4.3   Removal of nutrients in certain harvested crops (kg ha^{-1})*

Nutrient	Corn grain	Corn silage	Wheat grain	Soybeans	Coastal Bermuda grass	Tall fescue	Hardwood forest (annual uptake)
Nitrogen	123	185	80	134	448	280	94
Phosphorus	22	34	15	144	34	63	9
Potassium	28	168	16	40	186	280	29
Calcium	2	28	2	6	101	49	25
Magnesium	10	34	4	7	41	45	6
Yield**	3.52 m^3	56 000 kg	2.11 m^3	1.23 m^3	22 400 kg	15 700 kg	

*Transposed from Carlile and Phillips (1976)
**Per hectare

4.6.2 Assimilative capacity

The assimilative capacity is governed by the combination of the loss
mechanisms and a major problem is discerning their interaction.

   If crops are to be harvested, the overall capacity of the soil-crop
system to remove nitrogen can be roughly estimated as 150% of the
expected removal by harvest (Walcott and Cook, 1976); this approach
assumes that when leaching losses are minimal, the combination of
immobilisation and denitrification losses as a means of N removal
represents 50% of the crop N uptake. Hence the assimilative capacity is gauged by multiplying the appropriate data within table 4.3
by 1.5 (Carlile and Phillips, 1976). The approach has many weaknesses, one of which is that it takes little account of the particular
forms of nitrogen and their transformation.

   A modified approach described in Overcash and Pal (1979) for
assessing loading rates is based on matching the crop uptake rate to
the supply of available nitrogen as shown schematically in fig. 4.6.
In its essentials, the inorganic 'pool' comprises ammoniacal-N (not
lost by volatilisation), the waste nitrate and the nitrified fraction of
the organic-N. Account should also be taken of denitrification and

the slow release of available nitrogen from native soil organic-N (mainly bound in stable humus) together with organic-N immobilised from the waste. Keeney *et al.* (1975) incorporated these factors into a formula for determining the waste application rate:

$$\text{Waste application rate} = \frac{\text{Crop N requirement} - \text{residual available N in soil}}{\text{available N in waste}} \qquad [4.3]$$

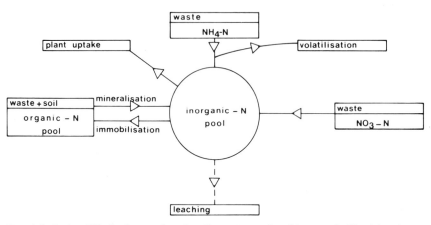

Fig. 4.6 A simplified scheme showing the net supply of inorganic-N arising from the composite forms of N within a waste

4.6.3 Municipal waste-water and sludges

In the land application of municipal waste-water and sludges, nitrogen is generally the land limiting constituent and the main technique for nitrogen management is via the crop N requirement (and harvesting) as indicated in equation [4.3].

By neglecting nitrogen fixation as an input and the changes in nitrogen storage in the soil (often important in the short-term, see Preul and Schroepfer, 1968), the long-term nitrogen balance is specified by:

$$N + cP = C + pQ + dN + vN \quad (\text{kg N ha}^{-1}\ \text{y}^{-1}) \qquad [4.4]$$

in which N is the total nitrogen in the waste; c, the concentration of nitrogen in P, the precipitation; C, the crop uptake; p, the permitted nitrogen concentration in the percolating or runoff water of volume Q; d, the fraction of N which is denitrified; and v, the fraction of N which is volatilised as ammonia.

The surface water balance is:

$$W + P = E_T + Q \qquad (\text{kg ha}^{-1}\,\text{y}^{-1}) \qquad\qquad [4.5]$$

in which W is the waste-water loading and E_T the potential evapotranspiration.

The nitrogen loading can also be expressed by.

$$N = yW \qquad\qquad [4.6]$$

in which y is the total nitrogen concentration in the waste-water. Combination of equations [4.4] to [4.6] leads to:

$$W = \frac{C + p(P - E_T) - cP}{y - p - y(d + v)} \qquad\qquad [4.7]$$

For design purposes the denitrification and volatilisation terms should be set at zero unless there is specific information at hand, this representing a worst-case situation. When equation [4.7] is solved on an annual basis, the crop removal is set equal to a constant for a given crop as shown in table 4.3. But crop requirements for nitrogen during the growing season approximately parallel the evapotranspiration demand; thus applications matching the seasonal changes in evapotranspiration may be more beneficial than a constant rate of application (see U.S. Environmental Protection Agency, 1977).

Equation [4.7] also serves as the base for determining liquid sludge applications and this aspect is discussed in Loehr et al. (1979). A slightly different approach based on equation [4.3] is detailed in Keeney et al. (1975).

Example 4.1 Nitrogen loading

A sludge containing 1.5% NH_4-N and 2.5% organic-N is distributed as a surface application to a crop for which the nitrogen requirement is 180 kg ha^{-1}. Soil analysis indicates that the residual available N derived from similar applications in previous years is 28 kg ha^{-1}. Estimate the application rate. It may be assumed that: (a) the applied sludge is not incorporated immediately into the soil and this reduces the available NH_4-N by a half (Keeney et al. 1975); and (b) that the mineralisation rate of the organic-N is likely to be 10% during the year under consideration.

Consider 1 kg sludge:

Available NH_4-N = 0.015 × 1 × 0.5 = 0.0075 kg
Available organic-N = 0.025 × 1 × 0.10 = 0.0025 kg
∴ Available nitrogen per kg of sludge = 0.0075 + 0.0025 = 0.01 kg

From equation [4.3]:

$$\text{sludge application rate} = \frac{180 - 28}{0.01} = 15\ 200 \text{ kg ha}^{-1}$$

For futher details see Keeney *et al.* (1975).

Example 4.2 Nitrogen balance in a slow rate infiltration system

A slow rate land infiltration scheme is designed on the basis of a long-term nitrogen balance using a forage crop as a growth medium, with a nitrogen uptake of 364 kg ha^{-1} y^{-1}. If the design precipitation is 1.28 m y^{-1} and the evapotranspiration 0.64 m y^{-1}, estimate the permitted hydraulic loading and the maximum nitrogen loading given that the applied waste-water contains 28 mg l^{-1} of nitrogen, and the maximum permissible nitrogen concentration in the percolate water is 10 mg l^{-1}. It may be assumed that denitrification and volatilisation accounts for 20% of the nitrogen load and that the concentration of nitrogen in the precipitation is negligible. (*Note*: 1 mg l^{-1} = 10^{-3} kg m^{-3}).

From equation [4.7]:
Hydraulic load (W) =
$$\frac{(364 \times 10^{-4} \text{ kg m}^{-2} \text{ y}^{-1}) + (10 \times 10^{-3} \text{ kg m}^{-3})(1.28 - 0.64)(\text{m y}^{-1})}{(28 - 10 - 28 \times 0.2) \times 10^{-3} (\text{kg m}^{-3})}$$
= 3.45 m y^{-1}

From equation [4.6]:

Nitrogen
load (N) $= y\ W = (28 \times 10^{-3} \text{ kg m}^{-3})(3.45 \text{ m y}^{-1}) \times 10^4 (\text{m}^2 \text{ ha}^{-1})$
= 966 kg ha^{-1} y^{-1}

For additional information see Loehr *et al.* (1979) and U.S. Environmental Agency (1977).

4.7 Phosphorus

Compounds containing phosphorus are present in almost every waste considered for land treatment. Organic forms of phosphorus are generally associated with microbial tissues, plant residues and the metabolic by-products of living organisms as in manures. Inorganic forms arise from detergent residues, fertilisers, etc. and are often more prevalent in sewage effluents.

Phosphorus, like nitrogen, is an essential nutrient for the growth of plants and animals and is of major importance in determining levels of eutrophication in surface waters.

Phosphorus in soils differs from nitrogen in that it cannot be lost in gaseous form. Hence phosphorus entering the soil either accumulates, or is removed, by: plant uptake and harvesting; soil erosion; or leached from the root zone in percolating waters.

4.7.1 Phosphorus retention

One of the unique characteristics of phosphorus is its immobility. Phosphorus immobilisation in the soil is related to the behaviour of the orthophosphate anions $H_2PO_3^-$ and HPO_3^{2-} to which other forms may convert. Since soil particles are negatively charged it might appear that these ion forms could leach away like the mobile nitrate ions. But this is not the case because the phosphates react rapidly with the soil solids and are retained by specific adsorption and precipitation reactions (section 4.4). As a result, the amount of phosphorus in the soil water is low, and most of the phosphorus in the soil is 'fixed' in forms unavailable to plants or for leaching.

In alkaline soils the fixation is due to the precipitation of insoluble calcium compounds whereas in acid soils, phosphorus fixation is associated with the adsorption on hydrous oxides of iron and aluminium and other clay minerals, together with precipitates formed with soluble iron and aluminium. The relative significance of these retention mechanisms is strongly dependent on pH.

As indicated in table 4.2 the adsorption/precipitation of phosphorus often represents the greatest fraction of the soil-plant assimilative capacity, but the capacity is finite. Unfortunately there is no simple method for determining the amount of phosphorus that the soil can retain before saturation occurs. Some indication of the retention properties can be obtained by assessing the adsorption isotherm, which describes the relationship between the phosphorus held in the soil and that remaining in the soil solution (Enfield, 1977); this yields an estimate of the assimilative capacity defined in terms of μg P g^{-1} soil. With knowledge of the bulk density of soil and the usable soil profile, the assimilative capacity can be expressed in terms of the unit kg P ha^{-1}. This rating was used by Sneider and Erickson (1972) to classify Michigan soils for use in municipal waste-water irrigation, but at present is not recommended as a technique for site evaluation until more experience has been gained.

4.7.2 Plant uptake

Phosphorus ranks next to nitrogen in importance to the fertility of soils as a macronutrient for plants. Plants take up phosphorus as phosphate ions and assimilate these into various cellular constituents.

An indication of the phosphorus uptake by plants is shown in table 4.3. As with other nutrients, tree vegetation removes less phosphorus than grasses. Typically, the harvested portions of annual crops contain only 10% or less of the fertiliser phosphorus added during the season in which the crop is grown, but recoveries as high as 50-60% are possible (Russell, 1973). Recovery is low, not only because the soil reacts with the added phosphorus to make it less available, but also because plants absorb considerable amounts of phosphorus from soil supplies, including residues from previous years. Hence the total removal per year as a fraction of the total added per year is more important than the recovery of that added during the year the crop is grown (U.S. Environmental Protection agency, 1977).

4.7.3 Runoff and leaching

Because of the immobility of phosphorus in the soil, losses arising from leaching are generally small and are usually only observed after prolonged waste application (say $>$ 5 years; Overcash and Pal, 1979).

Because phosphorus accumulates initially in the surface layer (see Loehr *et al.* 1979), runoff transport of P can be observed from pristine soils, agricultural areas or land application sites. The mode of transport is in the solution phase as well as phosphorus attached to soil particles. Since erosion is the only way that fixed phosphorus is lost, its control is very important; this implies that surface runoff should be minimised. In this respect vegetation is important for erosion control (see section 3.4) and for the maintenance of infiltration (section 2.3).

4.7.4 Assimilative capacity

From the previous discussion it is apparent that in a well-managed system, the design application rate is based on the combination of crop uptake and the retention capacity of the soil. The chosen application rate may continue until the accumulative capacity is exceeded. Thereafter, the rate should be reduced to crop uptake levels on the assumption that phosphorus is the land limiting constituent.

4.8 Acids, bases and salts

This represents an important class of primarily mobile constituents which as a group affect soil fertility and require control measures to provide optimum growth conditions.

Acids and bases refer to either organic or inorganic compounds which affect soil pH. Salts refer to elements important in agricultural irrigation and usually focus on sodium, calcium, potassium and magnesium.

When applied to the land, they induce an initial reaction or response. In the case of acids and bases, a neutralisation reaction may be induced, dependent on the soil buffering capacity (the ability of the soil water system to retain its pH). Salts react with the soil through the mechanisms of ion exchange and precipitation; the soil has little capacity to retain most soluble salts.

After an initial response, organic acids and bases are likely to undergo biological degradation yielding an array of end products. Inorganic acids and bases behave as conservative and non-decomposable species in a similar manner to salts. As with salts, they will eventually migrate with the movement of water (applied waste-water and rainfall). In the long-term, if the accumulation capacity of the soil is exceeded, there must exist an ion balance between the input and the outflow.

The assimilative capacity centres on these aspects and is viewed by Overcash and Pal (1979) as comprising three stages:

(a) *Stage 1*: The initial emphasis in design should be based on controlling the initial soil reaction to avoid sudden changes which may cause damage to the soil plant system.

(b) *Stage 2*: Ensure that there are no long-term effects by correcting imbalances between constituents.

(c) *Stage 3*: Loadings should be adjusted to keep the anions and cations in the drainage water within acceptable limits.

Conceptually the management of the salts and the acid/base constituents is broadly similar, in that emphasis is placed on retaining control parameters (such as pH or salinity) to within the tolerance limits imposed by the soil-plant system. In this respect the role of plants appears to be less direct in determining the assimilative capacity than in the case of nitrogen, which depended on crop uptake. However, this does not diminish the importance of the required control measures to ensure a thriving vegetative cover. Discussion will centre on the management of salts to illustrate the principles involved.

4.8.1 Salts

It is often the case in irrigation practice that more salts are added to the land than are taken by the crop. In the absence of rainfall the salt concentration in the soil water increases due to water losses incurred by evapotranspiration and with repeated irrigation the salts accumulate in the soil. The presence of excess salts in the root zone hampers crop growth for a variety of reasons, e.g. osmotic effects which restrict the rate of entry of water into the plant and induce a water deficiency (see section 1.4). In regions where there is sufficient rainfall at some period of the year the salts will be leached. However, in drier climates, it is usually necessary to provide additional irrigation water to remove the salts from the root zone.

Salts, because of their ionic form and mobility in solution, are often measured as the electrical conductivity (E.C.) of the soil water extract. If crop yields are not to suffer from salinity, the electrical conductivity of the soil water must be kept below the appropriate level for the crop (Ayers, 1977).

The management of salt-containing industrial wastes is broadly similar to the practices evolved in crop irrigation and focuses on:
(a)   the relative balance between sodium and other cations; and
(b)   the total salt content.

The general design approach for (a) is to add additional cations to correct the imbalance; while for (b) it is to ensure adequate salt movement through the soil.

MANAGEMENT OF SODIUM

No irrigation scheme can succeed unless the soil profile remains permeable and this depends on the proportion of sodium in the total exchangeable cations held in the soil and on the concentration of soluble salts in the percolating water. Large concentrations of sodium relative to other cations in the soil solution allow the sodium ions to dominate the charge sites on the soil colloids (see fig. 4.4); this weakens the stability of the soil aggregates causing dispersion and swelling. The consequent reduction in the size of the pores through which the solution is percolating, reduces the permeability.

The actual proportion of sodium in the total exchangeable cations that causes these swelling and deflocculation processes depends on many soil factors and is usually gauged in terms of the exchangeable sodium percentage (E.S.P.); this denotes the proportion of the cation exchange capacity occupied by sodium. A figure of 15% exchangeable sodium is widely quoted as the critical content above which the soil structure becomes unstable (Russell, 1973).

The distribution of exchangeable cations held in the soil matrix depends on the distribution of cations in the soil solution with which they are in equilibrium. The sodium limitation in the soil solution is often expressed in terms of the sodium adsorption ratio (S.A.R.) or the adjusted sodium adsorption ratio (see Ayers and Westcot, 1976), the S.A.R. being defined by:

$$\text{S.A.R.} = \frac{[Na]}{\sqrt{([Ca] + [Mg])/2}} \qquad [4.8]$$

in which [ ] refers to the salt concentration in meq l^{-1}. The S.A.R. may be interpreted as a measure of the sodium imbalance in the industrial waste or soil solution. The relationship between the E.S.P. and S.A.R. is not straightforward and can be determined empirically (Richards, 1954). Russell (1973) comments that the E.S.P. of many irrigated soils is approximately equal to the S.A.R. of the solution in equilibrium with the soil. As a general rule, the S.A.R. of applied waste-water should not exceed 10 (U.S. Salinity Laboratory Staff, 1953).

A number of management strategies exist for meeting the above conditions and are discussed in Overcash and Pal (1979). An obvious alternative is to pre-treat the waste by additions of calcium or magnesium to adjust the S.A.R. In the approach discussed by Eaton (1966), sufficient calcium is added to overcome the losses arising from calcium precipitation and crop uptake, together with an addition based on keeping the sodium concentration at less than 70% of total cationic concentration.

Besides its effect on the soil structure, sodium is perhaps the most critical ion affecting crop growth. These effects depend on the E.S.P. which can be used to classify plant tolerances to excess sodium. Further details are given in Bernstein (1974).

MANAGEMENT OF SALINITY

The leaching requirement (L.R.) is the fraction of the total amount of irrigation water that must be passed through the soil in excess of the crop water needed to control salinity to a specified level.

A mass balance between the applied liquid and the drainage water defines the leaching requirement:

$$\text{L.R.} = \frac{D_d}{D_i} = \frac{EC_i}{EC_d} \qquad [4.9]$$

where D_d and D_i are the equivalent depths of the drainage and irrigation water respectively and EC_d, EC_i their counterparts expressing the concentration of mobile salts in terms of electrical conductivity.

Equation [4.9] may be elaborated to include the combination of annual rainfall and the evapotranspiration losses from the plant-soil system. Taking account of these factors and neglecting the salt content of the rainwater yields:

$$D_i = \frac{EC_d}{(EC_d - EC_i)} \times (CU - P_e) \qquad [4.10]$$

where CU is the total water requirement of a particular crop (see section 2.7.1.3) and P_e is the effective rainfall (see Dastane, 1974). P_e refers to the amount of rainwater available to the crop during the growing season and is defined by:

$$P_e = P_g - (R + E_n) \qquad [4.11]$$

where P_g is the annual rainfall, R the annual runoff and E_n the evaporation from the soil surface in the non-growing season. In equation [4.10], EC_d may be set at the soil tolerance limit of the crop, such as listed in Ayers and Westcot (1976) and Ayers (1977).

Equation [4.10] may also be applied to particular mobile constituents simply by replacing the symbol EC by C, the constituent concentration. Considering a surface which is vegetated throughout the year and with runoff absent, yields the equivalence $CU - P_e \equiv E - P_g$ in which E signifies the annual evaporative loss; incorporating these changes into equation [4.10] leads to:

$$A = \frac{C_d - C_i}{C_d} \times \frac{Q}{E - P_g} \qquad [4.12]$$

where A is the land area requirement associated with the volume application rate Q ($= AD_i$) and C_i and C_d are the constituent concentrations in the irrigation and drainage water respectively. Inspection

Table 4.4 Impact of acidity and salinity on the growth environment

Physiological condition	Interpretation	Examples of plants suiting the quoted physiological condition category	References
pH of saturated soil paste:			
<4.2	Too acid for most crops to do well	Aspen	Overcash and Pal (1979)
4.2 – 5.2	Suitable for **acid-tolerant** crops	Wheat } } **slightly acid-tolerant** } }	U.S. E.P.A. (1977),
5.2 – 8.4	Suitable for most crops { **slightly acid-tolerant** } { **acid sensitive** }	Alfalfa } } **acid sensitive** }	Donovan et al. (1976)
>8.4	Too alkaline for most crops, indicates a possible sodium problem	Salt grass	
EC_e (mmho cm^{-1} at 25°C) of saturation extract:			
<2	No salinity problems		Loehr et al. (1979)
2 – 4	Restricts growth of **salt-sensitive** crops	Beans (field)	Ayers (1977)
4 – 8	Restricts growth of many crops	Alfalfa	
8 – 16	Restricts growth of all but **salt-tolerant** crops	Barley	
>16	Only a few very salt-tolerant crops make satisfactory yields		
E.S.P. (% of C.E.C.):			
2 – 10	Suitable for all except the **extremely sensitive** crops	Citrus	Loehr et al. (1979)
10 – 20	Affects **sensitive** crops; reduced permeability in fine soils	Beans	
20 – 40	Suitable for **moderately tolerant** crops; reduced permeability in coarse-textured soils	Tall fescue	
40 – 60	Only suitable for **tolerant** crops; permeability problems	Alfalfa, wheat	
>60	Unsuitable for all except the **most tolerant** crops; permeability problems	Tall wheat grass	

of equation [4.12] shows that the land area requirement is critically dependent on the selected value for C_d and highlights the controlling influence of rainfall and evapotranspiration.

CROP FACTORS

Crops vary in their tolerance to excess salts in the soil solution and tolerances vary with the stage of crop growth. A guide to the propensity of salinity problems is given in table 4.4. Ayers (1977) lists the expected yield reductions as a function of EC_e for a range of field, fruit, vegetable and forage crops. For example, tall fescue experiences no growth reduction at EC_e = 3.9 mmho cm^{-1}, but suffers a 50% reduction at EC_e = 13.3 mmho cm^{-1}.

4.8.2 Acids and bases

When the soil buffering capacity is exceeded, the soil pH can be altered via the application of an industrial waste. As a result many factors of the soil environment are changed simultaneously and affect the soil-plant assimilation of other constituents. For example, fig. 1.7 indicates that low and high pH preclude the availability of Ca, Mg and P which are essential nutrients. In the long-term the addition of acid or base constituents must be accompanied by other materials to neutralise these components. In addition to neutralisation, proper crop selection can enhance the assimilative capacity of soils. Acid-tolerant crops require less neutralisation of acid soil pH and this reduces pre-treatment costs. Forest trees seem to grow well over a wide range of soil pH values and are particularly tolerant of acid soils. For pasture grasses, many legumes and field crops, a range of pH from 5.8 to slightly above 7 is the most suitable (Brady, 1974). Usually when the soil pH is adjusted to the ranges shown in table 4.4, the effect of other nutrients can be optimised (see section 1.6).

Adjustment of pH can be achieved by adding limestone or sulphur to correct imbalances and is conceptually equivalent to the control of S.A.R. for salt imbalances. After neutralisation the salts produced must be evaluated for assimilative capacity in the manner described under salinity. Associated aspects are discussed in Overcash and Pal (1979).

Example 4.3 Salinity control

A crop to be irrigated has an annual water need (CU) of 750 mm and a salt tolerance limit (EC_d) taken as 9.0 mmho cm^{-1}. Climatic data indicates that annual rainfall (P_g) is 420 mm, the annual runoff (R) is 38 mm and the annual evaporation (E_n) during the non-growing season is 160 mm. Given that the irrigation water has a salt content of 1850 mg l^{-1}, estimate the irrigation water requirement (D_i).

From equation [4.11] the estimated effective rainfall during the growing season is:

$$P_e = 420 - (38 + 160) = 222 \text{ mm}$$

The correspondence between electrical conductivity and salt concentration is $(1 \text{ mmho cm}^{-1}) \doteq 1/640 \times (\text{mg } l^{-1})$ and implies $EC_i = 1850/640 = 2.90$ mmho cm^{-1}.

The salinity balance (equation [4.10]) yields:

$$D_i = \frac{9.0}{(9.0 - 2.9)} \times (750 - 222)$$

$$= 779 \text{ mm}$$

If the salt content of the irrigation water had been negligibly small, $D_i = CU - P_e = 528$ mm, indicating that 251 mm of drainage water is necessary for leaching excess salts to an acceptable level.

4.9 Metals

The metals category usually focuses on those having density $\gtrsim 5000$ kg m^{-3} which are arbitrarily designated as 'heavy metals'. Heavy metal wastes are generated by a number of mining, ore refining, metal-producing and electroplating industries. Municipal sludges generally contain heavy metals from domestic sources as well as industrial sources. However the concentration of metals in treated waste-water effluents is generally low, because most of the metals are transferred to the sludge during sewage treatment.

The common characteristic of these constituents is their tendency to accumulate in the upper region (~ 0.15 m) of the soil-plant system and only a small fraction of heavy metals is removed by plant uptake. Hence the assimilative capacity is usually determined from a critical soil level (see fig. 4.2). Beyond this, plant growth may be diminished. Where metals enter the edible portions of plants, food chain problems may occur. It is generally agreed that cadmium, zinc, copper, molybdenum and nickel are the elements in untreated waste-water most likely to be toxic to plants and/or animals. The element boron, though not a metal, may also be classed within this group (Loehr *et al*. 1979).

Effective removal of heavy metals from waste-waters depends on certain characteristic soil properties which must be taken into account when determining the loading rate of metal-carrying wastes. It should also be noted that virtually all the chemical elements in waste-water are normal components of the solid fraction of every soil sample. However elements in waste-water are usually in soluble form or become soluble during biodegradation (Overcash and Pal, 1979). In contrast many of the minerals in native soils are relatively insoluble, but are nevertheless important in the adsorption of heavy metals.

4.9.1 *Soil factors*

The uptake of metals by soils depend on a number of reactions which can be divided into two categories: (a) reactions with soil solid matter; and (b) reactions within the soil solution.

Reactions with soil solid matter include selective adsorption, covalent and ligand bonding and reversion processes.

Selective adsorption accounts for a small portion of the metals taken up by soil solids. However, such adsorption is not permanent, nor does it prevent some uptake by plant roots. Covalent and ligand bonding are mechanisms which incorporate ions into the organic matter of the soil and appear to be major factors in removing toxic elements. Reversion is a slow process in which ions are immobilised by substitution in the crystal lattice of the soil solids and thereby become inaccessible to the soil solution or plant uptake. Loehr *et al*. (1979) note that reversion probably accounts for no more than 10-20% of the total adsorption capacity associated with the mineral portion of the soil.

Interaction within the soil solution includes precipitation and methylation. Precipitation losses due to the formation of insoluble hydroxides are most significant in highly alkaline soils. Methylation refers to a reaction in which the hydrogen ion of the hydroxyl group (OH^-) of methyl alcohol is replaced by a metal and is associated with the metabolism of soil bacteria. Methylated elements are often both soluble and very toxic (especially to higher animals); they form a hazard that must be evaluated in land application schemes.

4.9.2 *Phytotoxicity*

Phytotoxic effects of metals and plant tolerance depend on the type and amount of metal in the solution phase and on soil factors such as pH, C.E.C. and soil organic matter.

In the classic dose-response curve (fig. 1.6) toxicity occurs when the substrate concentration (nutrient solution, soil, etc.) is sufficiently high. Table 4.5 indicates the lower and upper critical concentrations for Cd, Cu, Ni, Pb and Zn.

Table 4.5   Critical concentrations of elements in plant tissue*
(mg kg^{-1} dry matter)

	Cd	*Cu*	*Ni*	*Pb*	*Zn*
Background level	<0.5	8	2	3	40
Lower critical levels					
Deficiency threshold (plants)	–	4	–	–	20
Deficiency threshold (animals)	–	10	–	–	50
Upper critical levels					
Phytotoxic threshold	8	20	11	35	200
Zootoxic threshold	0.1–1 (background level)	30	50	~5	500

*Tentative data, generalised from various sources

Of the soil factors, studies have shown that soils with a high C.E.C. absorb more metals before levels toxic to the plant are reached than soils with low C.E.C. (McCormack, 1975). However, it has been recommended that the total application of heavy metals should not exceed 5% of the C.E.C. (Agricultural Research Service, 1974).

McCormack (1975) considers that soil pH is the principal determinant of the toxity of specific levels of heavy metals to plants. Soils with low pH have a low adsorption capacity for heavy metals; thus a metal content safe at a soil pH of 7.0 may be lethal to plants at 4.4, and calcareous soils are safest. Lindsay (1973) estimated that for every unit increase in pH there is a 100-fold decrease in the availability of zinc, cadmium, nickel and copper. Hence the control of pH is an important means of manipulating the soil to reduce the transfer of heavy metals to plants.

It is also well established that the toxic effects on plants are reduced by a high content of organic matter in the soil (Department of the Environment, 1972).

Threshold standards have been proposed for preventing phytotoxic build-up. One of the early proposals which has been widely used is the concept of the zinc equivalent (Chumbley, 1971) which states that the soil should have a pH > 6.5 and that the maximum metal addition to the soil should not exceed 250 p.p.m. of zinc or the zinc equivalent (Z.E.) defined by:

$$Z.E. \ (\mu g \ g^{-1}) = Zn^{2+} \ (\mu g \ g^{-1}) + 2Cu^{2+} \ (\mu g \ g^{-1}) + 8Ni^{2+} \ (\mu g \ g^{-1})$$
[4.13]

in which the unit $\mu g \ g^{-1}$ is on a dry solids basis. In equation [4.13] the following assumptions apply: (a) phytoxicity results mainly from zinc, nickel and copper in sewage sludges; (b) copper and nickel are about 2 times and 8 times more toxic than zinc respectively; (c) the toxicities of the three metals are additive. Assumptions (b) and (c) are open to question (Loehr et al. 1979); Leeper (1972) also noted that the formula lacked any factor to account for the capacity of individual soils.

Table 4.6   Maximum sludge metal addition to farmland (over the site lifetime)

Metal	Maximum metal addition (kg ha^{-1}) to soil with a C.E.C. (meq per 100 g) of:		
	< 5	5–15	> 15
Pb	500	1000	2000
Zn	250	500	1000
Cu	125	250	500
Ni	50	100	200
Cd	5	10	20

Developments of the zinc equivalent concept are considered in Chaney (1974) and Keeney et al. (1975). Table 4.6 summarises the U.S. Department of Agriculture guidelines which are discussed further in Dowdy et al. (1976).

4.9.3 Food chain hazard

Food chain problems arise from the uptake of toxic elements by plant roots, their transfer into the edible portions of the plants and their subsequent ingestion by primary and secondary consumers. The ensuing question concerns whether the elements being transferred are in sufficient quantities to present acute toxicity hazards to the consumers.

Generally plants represent an excellent barrier to this form of transfer for a variety of reasons. For lead, nickel and copper, the root provides the barrier since uptake and translocation are low. In many instances the food chain is protected because toxic elements accumulate preferentially in leaves, stalks and roots and less commonly in the seeds and fruit. With metals such as nickel and copper (see table 4.5), phytotoxicity provides a reliable barrier in that the plant dies or fails to grow long before it can accumulate a metal content toxic to mammalian consumers (Loehr *et al.* 1979). The factor of phytotoxicity is particularly significant in the control of cadmium in the food chain. For example, maintaining a Cd/Zn ratio of 1:100 in sludges or irrigation water, heavy metal accumulation of Cd and Zn in the soil will lead to phytotoxic conditions due to Zn before Cd concentrations can become detrimental to the food chain.

An exception to the plant barrier 'rule' (U.S. Environmental Protection Agency, 1977) is molybdenum which affects ruminants and is related to copper toxicity; molybdenum toxicity has been a serious practical problem to livestock producers in the western United States (Loehr *et al.* 1979).

Cadmium is currently the element of greatest concern as a food chain hazard to humans and has been studied extensively, e.g. Davis and Coker (1980).

An indication of the maximum cumulative cadmium application as a function of the soil C.E.C. is given in table 4.6. However this represents only a partial indication of the more detailed guidelines which are proposed in U.S. Environmental Protection Agency (1976) and elsewhere.

4.9.4 *Plant uptake*

Data reported in Spyridakis and Welch (1976) suggest that only a small fraction ($\lesssim 5\%$) of the applied heavy metals in waste-water is removed by plant uptake. Nevertheless there is considerable variation. Sidle *et al.* (1976) considered that in low heavy metal application areas, crop removal could significantly increase the irrigate disposal life of a site.

While some plants are accumulators of heavy metals, most crop plants are not. Davis and Carlton-Smith (1980) prepared a 'league' table for 39 crops showing their comparative efficiency of assimilating heavy metals from soils. These tables indicate which crops present are likely to pose the greatest hazard when grown on contaminated soil; they provide a basis for selecting indicator crops for particular purposes. The study indicated the importance of the plant genotype in discerning the significance of the zinc equivalent concept; it also confirms that soil concentrations of elements are not always closely associated with their likely toxicity to crops.

Plant species and even varieties differ widely in their tolerance to toxic metals. Vegetal crops that are very sensitive to toxic metals include members of the beet family, turnip, kale, mustard, lettuce and tomatoes. Beans, cabbage, collards and other vegetables are less sensitive. Field crops such as corn, small grains and soybeans are moderately tolerant. Most grasses (fescue, lovegrass, Bermuda grass, perennial rye grass) are tolerant to high amounts of metals (Epstein and Chaney, 1978).

Example 4.4 Metals loading

A sludge with a metals content of Zn = 5300 p.p.m., Cu = 1300 p.p.m., Ni = 450 p.p.m. and Cd = 100 p.p.m. (each on a dry weight basis) is applied to soil with C.E.C. = 10 meq per 100 g soil. An initial appraisal suggests that the metal loading per year might be based on the nitrogen fertiliser rate which is equivalent to 15 000 kg sludge (dry weight) ha^{-1} y^{-1}. Recent guidelines indicate that the permitted metal loading is equivalent to the amount of zinc needed to equal 5% of the soil C.E.C. and subject to an annual Cd loading limited to 2 kg ha^{-1}, with the total lifetime maximum set at 10 kg Cd ha^{-1}. Check the management strategy and estimate the lifetime of the site.

Estimate of zinc equivalent, Z.E., from equation [4.13]:

$\qquad$ Z.E. $\mu g/g$ = 5300 $\mu g/g$ + 2 (1300 $\mu g/g$) + 8 (450 $\mu g/g$)
$\qquad\qquad$ = 11 500 $\mu g/g$ or 11 500 $\times$ 10^{-6} kg Zn^{2+} kg^{-1} sludge

Condition on permitted metal loading:

$\qquad$ Atomic weight Zn = 66
$\qquad$ Equivalent weight Zn^{2+} = 66/2 = 33 g
$\qquad$ 10 meq Zn^{2+} = 10 $\times$ 10^{-3} $\times$ 33 = 0.33 g
$\qquad$ 5% of 10 meq = 0.0165 g Zn^{2+}
$\qquad$ i.e. 5% of soil C.E.C. (in Zn^{2+}) = 0.0165 g Zn^{2+} per 100 g soil
$\qquad\qquad\qquad\qquad\qquad$ = 165 $\times$ 10^{-6} kg Zn^{2+} per kg soil

Assume the metal is distributed over a 0.15 m depth of soil with bulk density 1250 kg m^{-3}, then the equivalent soil mass is 1250 $\times$ 0.15 $\times$ 10^4 = 1.875 $\times$ 10^6 kg ha^{-1}.

Permitted loading = $(165 \times 10^{-6}$ kg Zn^{2+} per kg soil$) \times (1.875 \times 10^6$
kg soil $ha^{-1})$
= 309 kg Zn^{2+} ha^{-1} (0.15 m depth)

If sludge loading is based on the nitrogen balance then:

Equivalent metal loading = $(1.5 \times 10^4$ kg sludge ha^{-1} $y^{-1}) \times$
(0.0115 kg Zn^{2+} per kg sludge)
= 172.5 kg Zn^{2+} ha^{-1} y^{-1}

Site lifetime (based on Zn^{2+}) = $\dfrac{309 \text{ kg ha}^{-1}}{172.5 \text{ kg ha}^{-1} \text{ y}^{-1}}$

= 1.8 y

Cadmium loading:

Annual cadmium loading limit = 2 kg ha^{-1}
Total cadmium loading = 10 kg ha^{-1}
Actual Cd loading = $(1.5 \times 10^4$ kg sludge ha^{-1} $y^{-1}) \times (100 \times 10^{-6}$
kg Cd per kg sludge)
= 1.5 kg Cd ha^{-1} y^{-1}
Site lifetime (based on Cd) = $\dfrac{10 \text{ kg Cd ha}^{-1}}{1.5 \text{ kg Cd ha}^{-1} \text{ y}^{-1}}$
= 6.7 y

Annual sludge loadings may be based on the crop nitrogen requirement, but particular attention should be paid to the cadmium
loading which is approaching the permitted limit. The site lifetime
is approximately two years and determined by the equivalent zinc
loading. This may be extended in a variety of ways, e.g. by reducing
the annual sludge loading and supplementing the nitrogen requirement with artificial fertilisers, or by means of increased tillage and
extending the depth through which the metals are distributed.

4.10 Aggregate bio-organic wastes

The assimilative capacity of a soil-plant system for biodegradable
organic matter, such as associated with food processing and municipal
wastes, is primarily dependent on the ability to maintain predominantly aerobic conditions for rapid microbial degradation. In addition
to the microbial oxygen demand, plant root systems must be sustained

under aerobic conditions to ensure adequate growth; the amount of air required varies widely with plant species. Adverse effects on agricultural crops have been detected when the partial pressure exerted by oxygen in the soil atmosphere is 10% of the atmospheric pressure; this is roughly half the partial pressure exerted by oxygen in the free atmosphere and corresponds to a vapour concentration of 140 mg $O_2\, l^{-1}$. For oxygen levels to be maintained, the biochemical oxygen demand must be balanced by supplies gained from the free atmosphere by diffusion through the upper soil layers. The supply rate depends on the concentration gradient and the oxygen diffusion coefficient; the latter depends on the soil character and the moisture content. Calculation procedures for assessing the oxygen budget and permitted loadings are outlined in Carlile and Phillips (1976).

Plant factors are generally secondary in determining the renovation capacity. Also, it should be recognised that, with judicious management, the soil-plant system has a considerable capacity for assimilating organic waste as indicated in table 4.1; this class of material is unlikely to be the land limiting constituent.

4.11 Crop selection

The selection of a site and vegetation type is moulded by economic factors in which the usual objective is to minimise the costs of the waste management system. At one extreme, the most viable option may be to integrate the waste disposal into an existing crop management system, e.g. by disposing of sludges on agricultural land. In this case, the designer has little choice in the selection of a suitable vegetation type except by choosing an alternative disposal site. At the other extreme, the treatment system may be largely independent of existing land use; here the designer has greater freedom in the selection of specific plant species consistent with the economic objectives and the environmental constraints.

Sopper (1973) enumerated the following criteria which need to be considered when selecting the vegetative cover for a land application site:

(a) water requirement and tolerance;
(b) nutrient requirement and tolerance;
(c) nutrient utilisation and renovation efficiency;
(d) sensitivity to potentially toxic elements and salts;
(e) insect and disease problems;
(f) season of growth and dormancy requirement;

(g) natural range;

(h) ecosystem stability;

(i) demand or market for the product.

Certain of these factors have already been discussed and others are considered in Loehr *et al.* (1979) and D'Itri (1982). In addition, it is imperative that crop selection should take account of the specific characteristic of the waste to be applied and of the soils involved.

Although agricultural land is most often used for land application, an alternative site may be necessary. Forest land may be the only choice in some regions perhaps because of its proximity, its abundance or the lower cost of site acquisition. In arid regions there has also been the practice of applying effluent to landscape areas such as golf courses, parks and recreational areas. There is also a great interest in the use of waste-water and sludge as a soil conditioner, and as a fertiliser substitute on drastically disturbed land – each facilitating their revegetation (Sopper and Kerr, 1979; Loehr *et al.* 1979). If sewage effluent is used in irrigated agriculture, particular attention must be given to the problems posed by pathogens and to undesirable plant responses, such as reviewed in Baier and Fryer (1973) and Aikman (1983).

Of the many land treatment systems in existence, Overcash and Pal (1979) note that there are only two or possibly three predominant vegetation types used, these being grass species, forest, and fibre or non-food crops. However this does not preclude the many other types of vegetation commonly associated with land application.

Grasses offer the greatest potential for nutrient uptake, particularly for nitrogen. Sod crops are also very effective in reducing erosion. Though grasses enhance filtration, grassed slow-rate systems allow for long periods of interaction between the waste constituents and the soil-plant system. Many of the successful forages used in treatment systems are fairly water-tolerant and are also relatively tolerant to high concentrations of dissolved solids and boron in waste-water; these include reed canary grass, coastal Bermuda grass, tall fescue and perennial rye grass. Their major disadvantage is their direct link with the food chain since nearly the whole plant is consumed in grazing or as hay. The potential for harvesting nutrients with perennial crops is generally greater than with annuals since annuals use less of the growing season for nutrient uptake.

A predominant advantage of forests is that they reduce the chance for food chain transfers. Undisturbed forests generally possess a high rate of infiltration, though unamended forest may be poorly drained compared to agricultural land; this may prove to be a limiting feature, depending on the water tolerance of the species. The uptake of nutrients by forests may be considerable during the growing season; much of these are redeposited through leaf fall, rather than being removed as in the case of harvested agronomic crops. Although harvesting woods removes accumulated nutrients, the long intervals between harvest makes nutrient removal low compared with the grasses. Nitrogen loading is often the limiting factor in using forest lands for waste treatment sites (Loehr *et al.* 1979). Compared to crop land there is a general lack of information concerning the effects of applying wastes to forest lands; much of the current experience is reviewed in Sopper and Kerr (1979).

Non-food or fibre crops likely to be important in land treatment systems include cotton, jute or hemp, and crops grown for the fermentation production of fuels (Overcash and Pal, 1979). These crops are annuals grown under managed conditions and pose the least threat to the food chain.

4.12 References

Agricultural Research Service (1974) 'Recommendations for management of potentially toxic elements in agricultural and municipal wastes'. *Factors Involved in Land Application of Agricultural and Municipal Wastes.* Washington, D.C.: Agricultural Research Service, U.S. Department of Agriculture.

Aikman, D.I. (1983) 'Waste-water reuse from the standpoint of irrigated agriculture'. *The Public Health Engineer* **11**(1), 35-41.

Anon (1973) 'The big farm at the end of the sewer'. *Farmers Weekly* **78** (part 18), page xi.

Ayers, R.S. (1977) 'Quality of water for irrigation'. *Proceedings of the American Society of Civil Engineers* **103** (IR2), 135-54.

Ayers, R.S. and Westcot, D.W. (1976) *Water Quality for Agriculture.* F.A.O. Irrigation and Drainage Paper No. 29. Rome: Food and Agriculture Organisation of the United Nations.

Baier, D.C. and Fryer, W.B. (1973) 'Undesirable plant responses with sewage irrigation'. *Proceedings of the American Society of Civil Engineers* **99** (IR2), 133-41.

Bernstein, L. (1974) 'Crop growth and salinity'. *Drainage for Agriculture.* Agronomy No. 17. Madison, Wisconsin: American Society of Agronomy, 39-54.

Brady, N.D. (1974) *The Nature and Properties of Soils.* 8th edn. New York: Macmillan.

Carlile, B.L. and Phillips, J.A. (1976) *Evaluation of Soil Systems for Land*

Disposal of Industrial and Municipal Effluents. Report 118. Raleigh, North Carolina: Water Resources Research Institute, University of North Carolina.

Chaney, R.L. (1974) 'Recommendations for management of potentially toxic elements in agricultural and municipal wastes'. *Factors Involved in Land Application of Agricultural and Municipal Wastes (Draft)*. National Program Staff, Beltsville, Maryland: Soil, Water and Air Sciences, Agricultural Research Service, U.S. Department of Agriculture, 97-120.

Chumbley, C.G. (1971) *Permissible Levels of Toxic Metals in Sewage Use on Land*. Advisory Paper No. 10. London: Agricultural Development Advisory Service, Ministry of Agriculture, Fisheries and Food.

Culp, G., Williams, R. and Lineck, T. (1978) 'Costs of land application competitive with conventional systems'. *Water and Sewage Works* 125 (October), 49-53.

Dastane, N.G. (1974) *Effective rainfall in irrigated agriculture*. F.A.O. Irrigation and Drainage Paper No. 25. Rome: Food and Agriculture Organisation of the United Nations.

Davis, R.D. and Carlton-Smith, C. (1980) *Crops as Indicators of the Significance of Contamination of Soil by Heavy Metals*. Technical Report TR 140. Stevenage, U.K.: Water Research Centre.

Davis, R.D. and Coker, E.G. (1980) *Cadmium in Agriculture, with Special Reference to the Utilisation of Sewage Sludge on Land*. Technical Report TR 139. Stevenage, U.K.: Water Research Centre.

Department of the Environment (1972) 'Agricultural use of sewage sludge'. *Notes on Water Pollution* 57 (June).

D'Itri, F.M. (1982) *Land Treatment of Municipal Wastewater: Vegetation Selection and Management*. Michigan: Ann Arbor Science.

Donovan, R.P., Felder, R.M. and Rogers, H.H. (1976) *Vegetative Stabilization of Mineral Waste Heaps*. Report No. EPA-600/2-76-087. Washington, D.C.: U.S. Environmental Protection Agency.

Dowdy, R.H., Larson, R.E. and Epstein, E. (1976) 'Sewage sludge and effluent use in agriculture'. *Land Application of Waste Materials*. Ankeny, Iowa: Soil Conservation Society of America.

Eaton, F.M. (1966) 'Total salt and water quality appraisal'. In Chapman, H.D. ed. *Diagnostic Criteria for Plants and Soils*. California: University of California, Division of Agriculture Science, 510-32.

Enfield, C.G. (1977) 'Wastewater phosphorus removal using land application'. *Civil Engineering-ASCE* 47 (June), 58-60.

Epstein, E. and Chaney, R.L. (1978) 'Land disposal of toxic substances and water-related problems'. *Journal of the Water Pollution Control Federation* 50, 2037-41.

Ives, K.J. and Gregory, J. (1967) 'Basic concepts of filtration'. *Proceedings of the Society for Water Treatment and Examination* 16 (3), 147-69.

Keeney, D.R., Lee, K.W. and Walsh, L.M. (1975) *Guidelines for the Application of Wastewater Sludge to Agricultural Land in Wisconsin*. Technical Bulletin No. 88, Madison, Wisconsin: Department of Natural Resources.

Laak, R. (1970) 'Influence on domestic wastewater pretreatment on soil clogging'. *Journal of the Water Pollution Control Federation* **42**, 1495-500.

Leeper, G.W. (1972) *Reactions of Heavy Metals with Soils with Special Regard to their Application in Sewage Wastes*. United States: Department of the Army, Corps of Engineers, under Contract No. DA CW73-73-C-0026.

Lindsay, W.L. (1973) 'Inorganic reactions of sewage wastes with soils'. *Proceedings of the Joint Conference on Recycling Municipal Sludges and Effluents on Land*. Washington, D.C.: Environmental Protection Agency and U.S. Department of Agriculture.

Loehr, R.C., Jewell, W.J., Novak, J.D., Clarkson, W.W. and Friedman, G.S. (1979) *Land Application of Wastes*. Volumes 1 and 2. New York: Van Nostrand Reinhold.

McCormack, D.E. (1975) 'Soils and their potential for safe recycling of wastewater and sludges'. *Proceedings of the 2nd National Conference on Complete Water Reuse: Water's Interface with Energy, Air and Solids*. Chicago, Illinois: American Institute of Chemical Engineers and Environmental Protection Agency Technology Transfer, 629-36.

Overcash, M.R. and Pal, D. (1979) *Design of Land Treatment Systems for Industrial Wastes: Theory and Practice*. Michigan: Ann Arbor Science.

Overcash, M.R., Klose, H.C., Rock, D., Marshburn, R. and Pal, D. (1978) 'Pretreatment land application of textile plant wastes'. *Water – 1977*. American Institute of Chemical Engineers, Symposium Series No. 178, **78**, 163-75.

Pound, C.E., Crites, R.W. and Reed, S.C. (1978) 'Land treatment: present status, future prospects'. *Civil Engineering-ASCE* **48** (June), 98-102.

Preul, H.C. and Schroepfer, G.J. (1968) 'Travel of nitrogen in soils'. *Journal of the Water Pollution Control Federation* **40**(1), 30-48.

Rich, L.G. (1973) *Environmental Systems Engineering*. New York: McGraw-Hill.

Richards, L.A. (ed.) (1954) *Diagnosis and Improvements of Saline and Alkali Soils*. Agricultural Handbook No. 60. Washington, D.C.: U.S. Department of Agriculture.

Russell, E.W. (1973) *Soil Conditions and Plant Growth*. 10th edn. London: Longman.

Sidle, R.C., Hook, J.E. and Kardos, L.T. (1976) 'Heavy metals application and plant uptake in a land disposal system for waste water'. *Journal of Environmental Quality* **5**, 97-102.

Sneider, I.F. and Erickson, A.E. (1972) *Soil Limitations for Disposal of Municipal Wastewaters*. Research Report No. 195. Michigan: Agricultural Experiment Station East Lansing, Michigan State University.

Sopher, C.D. and Baird, J.V. (1978) *Soils and Soil Management*. Virginia: Reston.

Sopper, W.E. (1973) 'Crop selection and management alternatives – perennials'. *Recycling Municipal Sludges and Effluents on Land*. Washington, D.C.: National Association of State Universities and Land Grant Colleges, 143-54.

Sopper, W.E. (1976) 'Use of the soil-vegetation biosystem for wastewater recycling'. In Sanks, R.L. and Asano, T. *eds. Land Treatment and Disposal of Municipal and Industrial Wastewater.* Michigan: Ann Arbor Science, 17-43.

Sopper, W.E. and Kerr, S.N. *eds.* (1979) *Utilization of Municipal Sewage Effluent and Sludge on Forest and Disturbed Land.* Pennsylvania State University Press.

Sopper, W.E. and Richenderfer, J.L. (1979) 'Effects of municipal wastewater irrigation on the physical properties of the soil'. In Sopper, W.E. and Kerr, S.N. *eds. Utilisation of Municipal Sewage Effluent and Sludge on Forest and Disturbed Land.* The Pennsylvania State University Press, 179-95.

Spyridakis, D.E. and Welch, E.B. (1976) 'Treatment processes and environmental impacts of waste effluent disposal on land'. In Sanks, R.L. and Asano, T. *eds. Land Treatment and Disposal of Municipal and Industrial Wastewater.* Michigan: Ann Arbor Science, 45-83.

Uiga, A., Wallace, A.T. and Howells, D.H. (1976) 'Let's consider land treatment, not land disposal'. *Civil Engineering-ASCE* (March), 60-62.

U.S. Environmental Protection Agency (1976) *EPA Technical Bulletin on Municipal Sludge Management.*

U.S. Environmental Protection Agency (1977) *Process Design Manual for Land Treatment of Municipal Wastewater.* EPA 625/1-77-008. Washington, D.C.: U.S. Environmental Protection Agency (Environmental Research Information Center Technology Transfer, Office of Water Program Operations) and U.S. Army Corps of Engineers and U.S. Department of Agriculture.

U.S. Salinity Laboratory Staff (1953) *Diagnosis and Improvement of Saline and Alkaline Soils.* U.S. Department of Agriculture Handbook No. 60. Washington, D.C.: U.S. Government Printing Office.

U.S. Soil Conservation Service (1970) *Irrigation Water Requirements.* Technical Release No. 21. Washington, D.C.: U.S. Department of Agriculture.

van't Woudt, B.D. and Hagan, R.M. (1957) 'Crop responses at excessively high soil moisture levels'. In Luthin, J.N. *ed. Drainage of Agriculture Lands.* Agronomy No. 7, 579-611. Madison, Wisconsin: American Society of Agronomy.

Walcott, A.R. and Cook, R.L. (1976) 'Crop and system management for wastewater application to agricultural land'. In Knezek, B.D. and Miller, R.H. *eds. Application of Sludges and Wastewaters on Agricultural Land.* A multiregional document sponsored by North Central Regional Committee NC-118 and Western Regional Committee W-124.

5 Shelterbelts

5.1 Introduction

The planting of shelterbelts represents a straightforward means of improving the climatic conditions within a locality – a practice which has been pursued for many centuries. Records show that it was ordained by the Scottish Parliament in 1457 and later, around 1789, some German emigrants exploited the systematic planting of trees on the steppes of Russia to protect agricultural crops against fierce winds. In the U.S.A. the use of large-scale planting dates from the beginning of the nineteenth century; during the period 1934 to 1942 the Soil Conservation Service of the U.S. Department of Agriculture planted more than 13 000 km of windbreaks and shelterbelts. Much similar work has been pursued elsewhere.

The terms shelterbelts and windbreaks are often used interchangeably though a distinction can be made. A shelterbelt is sometimes taken to refer to a belt or block of trees and/or shrubs arranged for protecting fields from wind, whilst a windbreak, though achieving the same purpose, may be artificial and associated with the protection of buildings and gardens. However the distinction is loose and there appears to be little agreement on the precise definition of the terms. In the present context they will be treated as synonymous except for situations where the protective value of plantings is a consequence of factors other than those arising from direct reductions in the wind speed, e.g. the retention of air-borne pollutants. In these particular cases the term 'shelterbelt' would seem more appropriate.

Although the most obvious effect of a shelterbelt is the reduction of wind speed there are many related factors which combine to alter the microclimate and lead to general improvements in the environmental conditions for plants, animals and people. On arable farms, crop yields are likely to be increased through sheltering and, in these days of energy shortage, windbreaks serve a useful role in reducing heat losses from dwellings and glasshouses. In arid regions of the world with persistent winds, windbreaks perform an important function in checking the wind erosion of soils. In forestry, windbreaks enhance the viability of plantations of young trees; at a later stage they provide protective margins and internal wind-firm strips which enhance the stability of the overall forest stand. Shelterbelts also play a useful role in checking pollution in its many shapes and forms, e.g. the control of snowdrift, noise, fog, dust and salt spray.

From this brief account, it is apparent that although the idea of a windbreak or a shelterbelt is conceptually straightforward, there are many avenues which can be explored to elucidate their form and character. Indeed, as a subject, they command a vast literature. Although a little dated, the review by van Eimern *et al.* (1964) remains one of the most comprehensive and is a valuable source of information covering many aspects of shelterbelt performance. Caborn (1965) provides a discerning account of the planning and design of shelterbelts and windbreaks in practice and complements the more theoretical approach adopted in van Eimern *et al.* (1964).

5.2 Shelterbelts and air flow

5.2.1 Atmospheric wind
Winds are created by differences in atmospheric pressure arising from distributions of the type shown on the typical weather map. From the isobaric spacing it is possible to determine the geostrophic wind, V_G, the wind which occurs at the top of the planetary boundary layer. The actual height of the boundary layer is variable (~ 600 m) and represents the elevation at which surface factors such as temperature and ground roughness cease to affect the upper level wind. It is the magnitude of the geostrophic wind which largely determines the wind speed observed at the earth's surface.

As the ground is approached, the wind speed falls due to the surface friction and becomes zero at the ground itself. In the lowest 100 m or so of the atmosphere the wind varies logarithmically with height (see fig. 5.1):

$$u = \frac{u_*}{k} \ln \frac{z}{z_0} \qquad z > z_0 \qquad\qquad [5.1]$$

where u is the wind speed at height z, u_* the friction velocity, z_0 the roughness length and k is von Karman's constant (= 0.4). The logarithmic variation holds over the cited height range provided temperature effects are negligible (termed as 'neutral' conditions) and provided the roughness length is reasonably uniform over a fetch of 10-20 km.

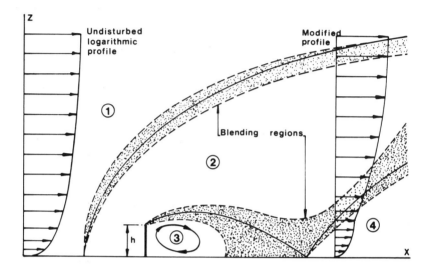

Fig. 5.1 The flow zones in a boundary layer disturbed by a shelterbelt: (1) undisturbed boundary layer flow; (2) region of influence of shelterbelt on pressure field; (3) region of flow separation (eddies); (4) re-establishment of boundary layer

Table 5.1   Roughness length (z_0) of natural surfaces

Surface	Height of roughness elements	z_0 (m)
Soils		$0.001 - 0.01$
Grass*	0.02–0.1 m	$0.003 - 0.01$
	0.25–1.0 m	$0.04 \ -0.1$
Agricultural crops*		$0.04 \ -0.20$
Forests*		$1.0 \ \ -6.0$

*z_0 depends on the wind speed

Typical values of the roughness length are shown in table 5.1 and a useful rule of thumb is to take z_0 as roughly $h/10$ where h is the average height of the roughness elements. Inspection of equation [5.1] shows that as the surface roughness increases, the wind speed at a given height is reduced due to the greater surface frictional force.

The friction velocity determines the scale of the wind speed near the surface and in neutral conditions can be adequately specified by the empirical function discussed in Counihan (1975) and Moore (1975):

$$u_* = \frac{V_G}{20} (1 + 0.24 \log_{10} \frac{z_0}{0.38})^{1/2} \qquad (z_0 \text{ in m}) \qquad [5.2]$$

which shows that u_* velocity increases with ground roughness. As a rough guide $u_* = 1/20 \times$ geostrophic wind speed, or over short vegetation is around $1/10 \times$ wind speed at 2 m above the ground.

5.2.2 Wind distribution

When wind encounters a shelterbelt or windbreak set perpendicular to the wind the air flow is disturbed due to the drag exerted by the belt and follows the pattern shown schematically in fig. 5.1. This indicates that the wind speed changes both vertically and horizontally. The presence of the obstacle causes an upward displacement of air and the air flow may be disturbed up to 3-4 times the height of the belt (H). Air subsides on the leeside of the belt due to the reduced pressure behind the barrier and subsequently regains energy until a distance of about 20H at which the vertical wind profile approaches the logarithmic form.

The greatest wind protection occurs roughly 2H behind the belt, the precise position depending on the porosity and the shape of the belt cross-section. As a general guide the sheltered zone starts about 5H to the windward and a reduction of 20% or more occurs within the distance 20H on the leeside (see fig. 5.2).

5.2.3 The significance of shelterbelt density (permeability)

Figure 5.2 shows that for dense belts the greatest wind reduction is very close to the belt and is furthest away for belts of medium density. The term density refers to the foliage area per unit volume but is usually specific in terms of the effective porosity of the belt. Experience suggests that optimum density (in terms of providing the maximum protected distance) corresponds to a 'porosity' of about 40-50%. This implies that when looking from the front, the mesh of leaves should cover 50-60% of the total frontal area. With the exception of snow control it is generally advantageous to ensure that the permeability, i.e. porosity, is evenly distributed throughout the height of the belt, perhaps being denser towards the ground (Caborn, 1957, page 36).

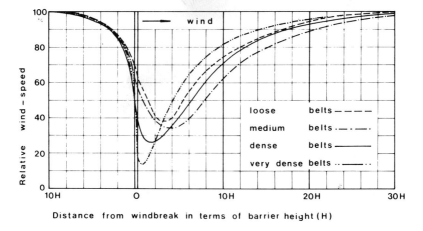

Fig. 5.2 Relative wind velocities in the vicinity of shelterbelts of different degrees of penetrability

5.2.4 *Effect of shelterbelt width*
The moderate degree of permeability implies the use of narrow belts. Single-row shelterbelts may be difficult to establish and maintain in good condition but have the advantage of saving on land and trees. Multiple-row belts are preferable, the width often being determined by the availability of land. Generally the higher the belt the more rows of trees required since there is a tendency for belts to become more open. Tanner and Nägeli (1947) suggest that low belts ($\sim$ 5 m) must be 2-5 m wide whereas high belts ($\sim$ 25 m) should be 10-15 m wide. If windbreaks are too wide there is a tendency for the sheltered area to be reduced. Caborn (1957, page 126) suggests that the width:height ratio above which the width becomes a limiting factor is about 5, though this depends on the porosity of the belt.

5.2.5 *Shelter in an oblique wind*
The effect of the belt is greatest when the wind blows perpendicular to it. As the angle between the wind and the belt is reduced the protected zone dwindles for obvious geometrical reasons. However, the effective porosity is also altered and a purely geometrical description is inadequate. Nevertheless for order of magnitude purposes a simple geometric view is quite useful. Figure 5.3 from Seginer (1975), illustrates some experimental data which shows that the relative protected distance varies linearly with the cosine of the angle of incidence α (measured from the normal to the shelterbelt). Though this view was not supported in the subsequent experiments of Seginer (1975), it suggests that the downwind distance $d(x)$ over

which the wind is reduced by at least 20% can be estimated by:

$$d(x) \sim 17H \cos \alpha \qquad\qquad [5.3]$$

which is a simplified version of the expression proffered in Woodruff and Zingg (1952). Skidmore and Hagen (1970, 1973) provide an alternative description of the wind speed reduction patterns measured leeward of a 40% porous barrier with data fitted by the expression:

$$\frac{u(x)}{u_0} = 0.85 - 4 \exp(-0.2H') + \exp(0.3H) + 0.0002 \, H'^2 \quad [5.4]$$

in which

$$H' = \frac{x}{\sin \theta}$$

where x is the leeward distance in barrier heights, $\theta = 90-\alpha$, the acute angle between the wind direction and the barrier direction, $u(x)$ the wind speed at the relative distance x and u_0 the open field wind speed. A minimum $\sin \theta$ was set at 0.18. On the windward side of the barrier the wind speed reduction was specified by:

$$\frac{u(x)}{u_0} = 0.502 + 0.197x - 0.019x^2 \qquad\qquad [5.5]$$

and is treated as being independent for wind angle.

Overall the value of considering the effects of wind angle is that it paves the way for estimating the minimum length required to obviate the effects of the belt extremities on centre-line protection. It is also useful when assessing barrier orientation and inter-barrier spacing when making use of windbreaks for combating soil erosion (see Skidmore and Woodruff, 1968).

5.2.6 Effect of shelterbelt cross-sectional profile

The 'ideal' cross-sectional profile of shelterbelts has been examined in many wind tunnel studies, e.g. Woodruff and Zingg (1952) and Caborn (1957) and depends on their desired purpose. An inclined windward edge diverts the air upward and reduces the effective drag encountered by the normal flow; this reduces the effective degree of penetrability and is of advantage in the marginal protection of forests (Caborn, 1957, page 126). On the other hand, belts with a more or less abrupt margin as shown in fig. 5.4, allowing air to filter through the belt, are more suitable for providing shelter near the ground; this may be more suitable for agricultural requirements.

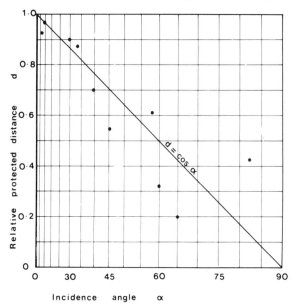

Fig. 5.3 The relative protected distance in the lee of a windbreak related to the incidence angle of the wind

Fig. 5.4 Belts with a 'pitched-roof' cross-section, (a), are less effective in halting winds than belts with more or less vertical edges, (b)

5.2.7 *Effect of shelterbelt length*
The length of a belt is important when considering 'end' effects as evident in fig. 5.5. For a belt set perpendicular to the wind the central region downwind of the belt ceases to be affected by the extremities of the belt provided its length is about 12H, perhaps being extended to 24H for a wind which impinges on the belt at 45° (Bates, 1944; see also Example 5.1).

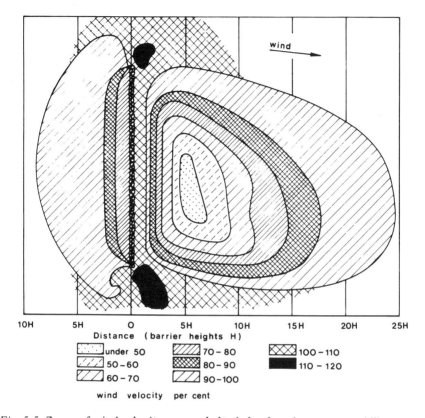

10H   5H   O   5H   10H   15H   20H   25H
                    Distance (barrier heights H)

under 50   70 – 80   100 – 110
50 – 60   80 – 90   110 – 120
60 – 70   90 – 100

wind   velocity   per cent

Fig. 5.5 Zones of wind velocity near a shelterbelt of moderate permeability

Example 5.1 Windbreak length in an oblique wind

For wind blowing perpendicular to a belt it is found that an area located at a distance $17H$ downwind of the centre of the belt is adequately protected provided the belt length is about $12H$. If the wind direction is at an angle α to the perpendicular axis of the belt, estimate the length of the belt required to retain the equivalent protection within the region.

From fig. 5.6(a), corresponding to $\alpha = 0$, end effects become negligible at a distance $d \ (\doteqdot 17H)$ downwind of the centre when:

$$d = \frac{L}{2} \tan \theta \qquad\qquad\qquad [5.6]$$

where L is the belt length and θ the angle between the observation point (O) and the belt extremities (A, B).

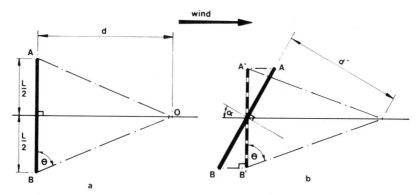

Fig. 5.6 Geometric model of windbreak protection in an oblique wind

For an oblique wind the projected length of the belt as seen by the wind is $A'B' = L \cos \alpha$ (fig. 5.6(b)). If it is tentatively assumed that the projected area of the belt behaves in the same way as the regime $\alpha = 0$ in so far as end effects are concerned, then the limit of protection can be specified generally by a point d' which lies at a distance d' from the belt defined by:

$$d' = \frac{L}{2} \cos^2 \alpha \tan \theta \qquad [5.7]$$

If the required length of the belt in an oblique wind is specified by L' consistent with the identity $d = d'$ then

$$L' = \frac{2d}{\cos^2 \alpha \tan \theta} = \frac{L}{\cos^2 \alpha} \qquad [5.8]$$

which by substituting $L = 12H$ yields:

$$L' = \frac{12H}{\cos^2 \alpha}$$

For example if $\alpha = 45°$, the equation implies that the required length of the belt must be $L' \sim 24H$.

5.2.8 *Effect of a gap in the shelterbelt*
It is not always possible to plant shelterbelts without gaps, e.g. resulting from road entrances. Gaps tend to promote jetting, i.e. excess wind velocities as illustrated in fig. 5.7 from Nägeli (1946). This shows that wind increase begins at a distance of $5H$ to the windward and continues to be noticeable up to $14H$ to $18H$ on the

leeside. Where a gap is necessary it should have an oblique alignment (see fig. 5.8) unless there is a danger of the entrance becoming blocked with snow (Caborn, 1965, page 202).

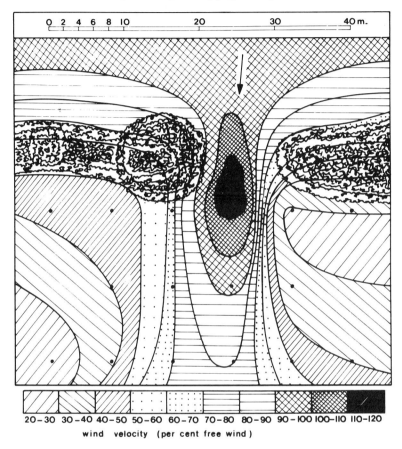

Fig. 5.7 Wind conditions in the vicinity of a gap in a windbreak

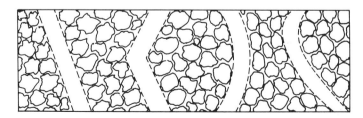

Fig. 5.8 Ways through cross-wind belts may be angled or curved to keep out the wind, but are liable to fill with drifted snow

5.2.9 Systems of shelterbelts and windbreaks

A single shelterbelt offers only limited wind reduction and for the protection of larger surfaces the erection of several belts is necessary. Examples of this type of planting are found in the dry windswept plateaux of the American prairies and the Russian steppes where the principal purpose is to control soil erosion. Plantings may exist as a system of parallel belts or as a mesh of crossing belts. The former are used where the main wind direction and the opposite wind direction greatly exceed all others, but if other winds also have a considerable influence the latter system provides better protection.

One of the major problems of extending the principles of the single shelterbelt, is that belts do not act independently unless they are a sufficient distance apart. Even in these circumstances, the effective roughness length of the terrain must be altered, as shown in the following section.

EFFECTIVE SURFACE ROUGHNESS

For a system of isolated roughness elements of height H, Lettau (1969) proposed the relationship:

$$z_0 = 0.5 \frac{HS}{A} \qquad [5.9]$$

where S is the average silhouette area (projected area on a plane normal to the flow) of the roughness elements, and A is the average area occupied by an element. In the case of a system of long parallel impermeable windbreaks of height H, equation [5.9] transforms to:

$$z_0 = 0.5 \frac{H^2}{L} \qquad [5.10]$$

where L is the spacing between the belts. When the belts are more closely spaced Businger (1975) argued that the effective height (aerodynamically) was no longer H but should be reduced, and proposed:

$$z_0 = 0.5 \frac{H^2}{L} (1 - \frac{CH}{L}) \qquad [5.11]$$

in which C is a coefficient (probably ~ 0.5) such that $CH/L \leqslant 1$. Unlike equation [5.9], Businger's formulation has not been substantiated, though some qualitative support was claimed in the study by

Iqbal *et al.* (1977).

By substituting the estimate of z_0 into equations [5.1] and [5.2] the effect of the average wind speed can be roughly gauged provided: (a) the system of belts is sufficiently extensive; and (b) that the reference wind speed is measured at a height several times greater than z_0; preferably this should exceed the shelterbelt height in order to avoid local effects arising from the nearest belt upwind. An analysis based on these principles is outlined in example 5.2.

PARALLEL BELTS: THE BEST INTERVAL?

There has been considerable interest in selecting the 'best' interval for a system of parallel belts. Because so many factors must be taken into account a full answer is not possible. In the first instance it depends on particular aims of the system. For example, the higher the prevalent wind speed and the more sensitive the soil to erosion, the smaller the interval must be. Other factors include the shape, density and height of the belts that form part of the system.

According to Kreutz (1958) a reduction of the inter-belt distance to below $10H$ affords practically no further protection and it also becomes uneconomic to use smaller spaces unless there is intensive use of the soil. At the other extreme there appears to be general agreement that belts should not be more than $30H$ to $40H$ apart, which corresponds to the maximum area which is likely to be protected by a single belt. Caborn (1957) indicates that if the maximum distance between the belts is $25H$ the intervening area will be sheltered to some extent. An indication of the expected wind reduction can be gleaned from reviews such as van Eimern *et al.* (1964). Otherwise one can make use of wind tunnel studies or faces the difficulty of trying to superimpose overall roughness effects on isolated belt behaviour. An *ad hoc* model which might be considered is to speculate that the minimum percentage reduction in wind speed at a reference height can be attributed to the effect of increasing the surface roughness of the terrain and can be estimated along the lines set out in example 5.2. Ideally the selected reference height should be sufficiently great to avoid local effects arising from any individual shelterbelt (say $z \gtrsim 3\,H$). If shelterbelts are less than $20H$ apart there should be a further percentage reduction in wind speed (say $\gtrsim 20\%$), this applying to the average velocity field between the belts. The logic behind this approach is that the modified wind speed occurring at the chosen reference height is treated as a free stream velocity and this scales the wind speed at lower heights.

Example 5.2 Effect of a system of shelterbelts on wind speed

During a study of the effects of a system of shelterbelts on the wind profile, Seguin and Gignoux (1974) compared the vertical wind profile over an open airfield characterised by the roughness length $z_0 = 0.013$ m and friction velocity $u_* = 0.63$ ms^{-1}, with equivalent profiles measured simultaneously within an area protected by shelterbelts, these being characterised by $z_0'' = 0.46$ m and $u_*'' = 0.83$ ms^{-1}.

It was observed that the average shelterbelt height $h = 8$ m and the inter-belt spacing/height ratio ~ 5. Verify whether the observations are consistent with the semi-empirical predictions of equations [5.2] and [5.11] and estimate the expected velocity reduction within the protected zone at a height of 12 m.

From equation [5.11] substituting $H/L = 1/5$ and $C = 0.5$, yields z_0'' (estimate) = 0.72 m which is high compared with the measured value $z_0'' = 0.46$ m.

Substitution into equation [5.2] yields the calculated ratio:

$$\frac{u_*}{u_*''} = \left[\frac{1 + 0.24 \log_{10} 0.013/0.38}{1 + 0.24 \log_{10} 0.72/0.38}\right]^{1/2} \qquad [5.12]$$

$$= 0.78$$

which is in good agreement with the measured ratio $0.63/0.83 = 0.76$.

From equation [5.1] the ratio of wind speeds at the height 12 m is given by the ratio:

$$\frac{u''}{u} = \left[\frac{u_*''}{u_*}\right] \frac{ln\ (12/0.72)}{ln\ (12/0.013)} \qquad [5.13]$$

$$= 1.28 \times 0.41$$
$$= 0.53$$

According to the direct measurements of Seguin and Gignoux taken at 12 m, $\bar{u}'' = 6.79$ m s^{-1} and $\bar{u} = 10.54$ m s^{-1} yielding $\bar{u}''/\bar{u} = 0.64$ which shows that the 'open' wind value is reduced by 36% within the protected area; this compares with an estimated reduction of 47%, the over-prediction caused by the over-estimate of z_0''.

Overall, it is seen that the semi-empirical predictions are broadly consistent with the measured data; and it is evident that the increased aerodynamic roughness — caused by the presence of shelterbelts — leads to a significant reduction of the average wind speed within the protected area.

5.3 Shelterbelts and wind erosion control

From the review in chapter 3 it is apparent that wind erosion is a serious problem in many parts of the world. Techniques for controlling the erosion are varied, but may be classified as short-term and long-term. Short-term measures include roughening the soil surface, increasing the percentage of erodible clods and a variety of agricultural practices outlined in Chepil and Woodruff (1963). More permanent measures include the maintenance of a vegetative cover or the planting of windbreaks, the latter having proved their value in many regions suffering erosion problems (van Eimern *et al.* 1964). One of the most extensive programmes of shelterbelt planting has been carried out in the Great Plains region of the United States with much of the recent experience reported in Great Plains Agricultural Council (1976).

The two main effects of shelterbelts which aid wind erosion control are: (a) they decrease the surface shear stress by reducing the wind speed; and (b) they trap moving soil. To make best use of windbreaks several characteristic parameters can be manipulated, including porosity, height, shape, spacing and orientation. Amongst these Hagen (1976) considered that windbreak porosity has most influence on wind speed reduction and soil trapping, the maximum wind and erosion reduction occurring when the windbreak porosity is about 40%. The height, of course, can be used to scale the dimensions of the protected area and this leaves orientation and belt spacing as the principal variables to be determined. However before discussing this further, it is useful to dwell on certain basic aspects of wind erosion.

5.3.1 Wind erosion forces

From the review in section 3.4.2 it was evident that the erosive potential of the wind is proportional to the wind speed cubed, provided the wind speed is above a threshold value. Estimates of the threshold value vary, but the value 5.4 m s^{-1} (12 m.p.h.) at a height of 1 m represents a convenient guide which has been adopted for design purposes in Skidmore and Woodruff (1968) when evaluating soil losses caused by wind erosion. It is also necessary to consider the distribution of wind speed both in terms of space and time. For example a 20 m s^{-1} wind of 1 hour's duration is equivalent to a 10 m s^{-1} wind of 8 hours' duration in terms of its erosive potential. This feature can be generalised by the expression fU^3 where f is the fraction of time wind speeds (U) are greater than the threshold.

From this Skidmore (1965) developed the concept of the wind erosion force vector, r_j, defined by:

$$r_j = \sum_{i=1}^{n} f_i \overline{U^3}_i \qquad\qquad [5.14]$$

where U_i is mean wind speed within the ith speed group and f_i is a duration factor which is expressed as the percentage of the total observations that occur within the ith speed. The subscript j indicates direction and takes the values 0 to 15 corresponding to the principal compass directions. The convention adopted by Skidmore (1965) and others, is to take a line pointing east as the initial side of the coordinate system (i.e. $r_{j=0}$ represents an erosion vector pointing east) and j increases in the anticlockwise direction (i.e. $r_{j=4}$ corresponds to an erosion vector pointing north, see fig. 5.9).

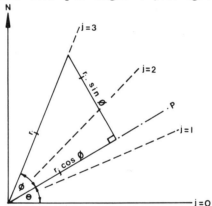

Fig. 5.9 Coordinate system for defining wind erosion force vectors

The sum of the magnitudes of the wind erosion force vectors for all directions gives the total magnitude of wind erosion forces for the given location:

$$F_T = \sum_{j=0}^{15} \sum_{i=1}^{n} f_{ij} \overline{U^3}_{ij} \qquad\qquad [5.15]$$

Equation [5.15] is a useful measure in that it scales the potential soil loss (see section 3.4.2).

The magnitude of erosion forces parallel and perpendicular to a selected direction can be obtained from the wind erosion force vectors. In fig. 5.9 if OP is the chosen direction at angle θ from the datum and, ϕ_j the angle between r_j and OP, the components of r_j parallel to OP may be defined by $r_j \cos \phi_j$, and perpendicular to OP by $r_j \sin \phi_j$. From this coordinate system it is evident that ϕ_j can be defined by:

$$\phi_j = j \times 22.5 - \theta \qquad [5.16]$$

Thus by taking the sum over the 16 principal compass directions the effective wind erosion force parallel to OP is:

$$F_{\|} = \sum_{j=0}^{15} r_j \, | \cos (j \times 22.5 - \theta) | \qquad [5.17]$$

All values of the trigonometric function are treated as positive because the erosion force along OP depends on its absolute value, i.e. it makes no difference whether the wind is in the direction OP or PO. Similarly the sum of the erosion forces perpendicular to OP is:

$$F_{\perp} = \sum_{j=0}^{15} r_j \, | \sin (j \times 22.5 - \theta) | \qquad [5.18]$$

5.3.2 Barrier orientation

When considering the orientation of barriers, one needs to consider the wind erosion forces occurring parallel to the barrier as well as those that occur along its normal. Since the direction of the minimum erosion forces is not necessarily perpendicular to the direction of the maximum erosion forces, it would be inadvisable simply to place barriers perpendicular to the direction at which $F_{\|}$ is a maximum, since this will not guarantee that $F_{\perp}$ is at its minimum. To resolve this situation Skidmore (1965) introduced the ratio:

$$R = \frac{F_{\|}}{F_{\perp}} \qquad [5.19]$$

The magnitude of R depends on the selected direction (θ) as illustrated in fig. 5.10. The orientation θ_R at which R is a maximum, viz. R_m, may be considered as the prevailing wind direction. The magnitude of R_m indicates the preponderance of wind erosion in the prevailing wind direction. The greater the value of R_m, the greater the prevalence of the prevailing wind erosion. A value $R_m = 1$ indicates no prevailing wind direction and a barrier is equally effective in any direction (see fig. 5.11), whereas $R_m = 2$ indicates a prevailing wind direction in which the erosion forces parallel to the prevailing wind are twice as great as those occurring perpendicular to this direction. Inspection of fig. 5.12 illustrates that there is a clear incentive to orient a barrier perpendicular to θ_R in order to maximise the area protected. It may not always be feasible to orient the barrier in its optimum direction and the consequences are examined in Skidmore and Woodruff (1968). One immediate remedy is to reduce the spacing between adjacent barriers.

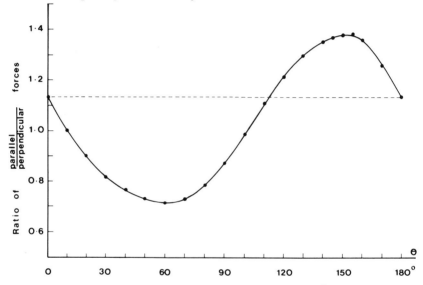

Fig. 5.10 Ratio of parallel to perpendicular forces as a function of bearing (see fig. 9)

Though orientation is a critical factor in windbreak design, it should be remembered that shelterbelts are just one weapon in the armoury of the soil conservation engineer; often the emphasis will be on the reduction of excessive wind erosion to some tolerable limit using a combination of agronomic practices. For this reason, and because of the variability of soil and climatic conditions, it is not possible to provide rigid rules for assessing the spacing of barriers. However if an acceptable level of soil erosion can be defined, then solution of the wind erosion equation (see section 3.4.2), provides a useful guide to evaluate field control measures and the effects of sheltering. Specific examples are discussed in Woodruff and Siddoway (1965) and Hagen (1976).

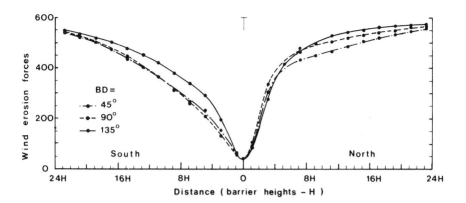

Fig. 5.11 Wind erosion forces at indicated distances perpendicular from 40% porous barrier when the barrier direction (B.D.) is 45°, 90° (east-west) and 135° respectively; wind data for Bismark, North Dakota

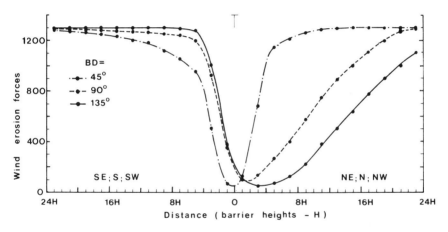

Fig. 5.12 Wind erosion forces at indicated distances perpendicular from 40% porous barrier when the barrier direction (B.D.) is 45°, 90° (east-west) and 135° respectively. Wind data are for Great Falls, Montana

5.4 Shelterbelts and microclimate

The shelter that lies behind a windbreak is more than just a drop in wind, it induces a series of changes in the microclimate, as indicated in fig. 5.13. This shows that when the wind speed is reduced, air, soil temperature and humidity tend to increase whereas evaporation losses are reduced. Away from the immediate vicinity of the belt the combination of factors generally leads to an increase in crop yield, this representing one of the major benefits of shelterbelts.

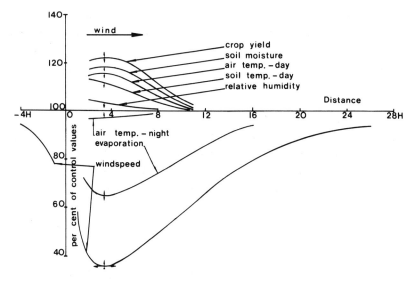

Fig. 5.13 Summary diagram of the effect of barriers on micrometeorological factors. The arrows indicate the directions in which values of different factors have been found to vary relative to the control values measured in unsheltered areas; H is the barrier height

5.4.1 Radiation

Rosenberg (1976) and others have indicated that the influence of shelterbelts on the radiation balance is negligible, except within a small area adjacent to the belt. During the day there may be increases or reductions in the radiant energy reaching the ground adjacent to the belt, depending on its orientation. Shadow length can be calculated from the following formula (see Caborn, 1965):

$$L_{sh} = H \sin (180 - \beta + \gamma) \cot \alpha \qquad [5.20]$$

where L_{sh} is the length of the shadow, H the shelter belt height, α the altitude of the sun, β the azimuth of the sun and γ the direction of the shelterbelt (with β and γ measured clockwise from north). At night, outgoing radiation may help to reduce nocturnal cooling but the effect is thought to be negligible except very close to the belt. Shading effects can depress crop growth but this may be offset by more favourable moisture conditions (see Marshall, 1967).

5.4.2 Air temperature

There can be marked increases in air temperatures close to barriers on the sun-facing side, the reverse happening on the side in shade.

During the day-time temperatures are likely to be higher over well-watered plants since the transport of sensible heat away from the crop is reduced (Rosenberg, 1976). At night, temperatures usually differ only slightly from the open in the presence of shelter (Marshall, 1967) and temperature inversions are less likely to be disrupted by turbulence.

5.4.3 Evaporation and transpiration

Evaporation is sometimes considered as one of the most important factors associated with shelter. Its reduction in sheltered areas allows conservation of soil moisture directly and indirectly through lowered transpiration. For an open water surface an increase in wind leads to a decline in the aerodynamic resistance (see section 2.2.4) and this always increases the evaporation rate. However this may be offset by changes in air temperature and humidity. For plants the transpiration rate need not be influenced by wind in the same way as evaporation from an open water surface. Using equation [2.13] Grace (1977) argued that the effect of wind on water use depended markedly on the physiological state of the plant. The analysis suggested that shelter will conserve water only where the crop is well-watered and there is little physiological stress (i.e. when the bulk physiological resistance is small). Much of the early work reviewed by van Eimern et al. (1964) indicates that savings of soil water do not always occur. Where savings have been observed it is sometimes because of the extra supplies derived from the melting of snowpacks built up near the windbreak.

5.5 Shelterbelts and snow management

The influence of shelterbelts on the control of snow cover is one of the most important aspects of shelterbelt use – particularly in climates with much snow or where snow represents a significant fraction of the gross precipitation. In many regions of Russia and North America, which suffer from dry summers, shelterbelts are used extensively for trapping snow over as wide a cropping area as possible – for soil water recharge and subsequent crop production. Belts also have the task of protecting roads and buildings from snow drifts (see section 5.8.3). Winter sports also benefit from the judicious use of plants and fences to redistribute snow in order to enhance ski and toboggan runs.

5.5.1 Snow distribution

The behaviour of blowing snow has much in common with sand or dust, an important difference arising from its lower density and evaporative character (Tabler, 1975). After snowfall reaches the ground, snow particles may be driven along by surface creep or saltation, depending on the wind speed, and this leads to a redistribution. The pattern of snow drifting behind or near a barrier reflects the local wind conditions, deposition occurring wherever there is a drop in wind speed. Nägeli (1953) indicated that the denser the belt, the higher and shorter the deposit on both sides of the belt; whereas with permeable belts the drifts are long and shallow. For dense belts the drift extends to a distance $2H$-$5H$ on the lee, which compares with the distance $15H$-$25H$ on the lee of permeable belts.

Windbreaks used for collecting snow should be dense or moderately dense and if trees are to be used they should be augmented by shrubs or ground cover as trees by themselves are not sufficient to initiate drifting. Caborn (1965) suggested that belts with an outer row of shrubs trap most of the snow within about 3-8 m on the leeward edge. Immediately near dense belts snow drifts can be larger to the windward than on the lee. Data reported in van Eimern *et al.* (1964) also shows that more snow collects within a dense belt than its permeable equivalent, with more snow collecting on the windward side. As a rule the density of belts increases with their width, and a five-row belt is said to catch most of the snow in itself (see van Eimern *et al.* 1964).

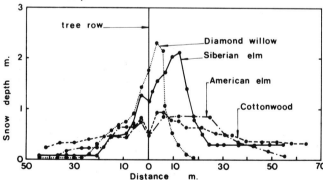

Fig. 5.14 Snowpack depth adjacent to single-row field windbreaks of diamond willow (6.7 m tall, 0.6 m spacing), Siberian elm (4.3 m tall, 0.9 m spacing), American elm (5.5 m tall, 1.8 m spacing) and cottonwood (18.3 m tall and 3.7 m spacing)

Where windbreaks are planned to spread snow as uniformly as possible, narrow belts that can be kept thin at the bottom are

preferable; this allows the snow to sweep through. Figure 5.14, from
Frank and George (1975), illustrates that the distribution of the
snowpack can be effectively controlled by species selection and
spacing distance. Data in fig. 5.15 emphasises that the permeability
of the lower section of the windbreak has a significant influence on
the distribution, but this may also affect the total snow volume
deposited (Frank and George, 1975).

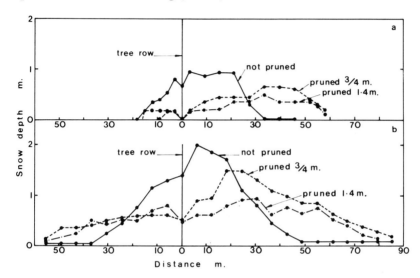

Fig. 5.15 Snowpack depth in (a) 1967, and (b) 1969, adjacent to test sections
of a 6.1 m single-row Siberian elm windbreak in its natural growth state and with
the bottom 0.8 and 1.4 m of branches removed

5.5.2 Snow cover management for water supplies
The particular value of snowmelt for sustaining water supply during
dry periods depends on whether the belt has been designed as a snow
'accumulator' or as a snow 'distributor'.

Moisture supplies gained from snow which has accumulated within
the belt can have a significant bearing on their health in dry summers.
For example the studies of Stoeckler *et al.* (1941) indicated that soil
water recharge from the snowpack was one of the main factors why,
in the northern Great Plains of America, trees live 20-30 years longer
and are more vigorous than those in the southern Great Plains. If a
belt designed for snowpack accumulation is too wide, George (1943)
observed that the leeward rows can suffer from lack of moisture
beyond the drift line compared with the windward rows. Patten
(1956) suggests that 5-8 rows of trees will usually be sufficient to act
as a good barrier to the wind, yet allow most of the snow to drift

inside the trees. Although there is an incentive to plant narrower windbreaks (say 3-5 rows) so that trees occupy less crop land it was pointed out by Frank and George (1975) that these might be inadequate in the more extreme weather conditions.

Where the objective is to trap snow over a wide cropping area shelterbelts must be relatively permeable at the bottom (see fig. 5.15) and single-row windbreaks are often employed. Besides shelterbelts, many studies, e.g. Black and Siddoway (1971, 1975), have shown the benefit of retaining stubble and narrow strip barrier systems, e.g. tall wheat grass as a means of increasing snow retention and reducing wind erosion.

5.6 Planting and noise reduction

Excessive noise is one of the most widespread social irritants and is particularly prevalent in the urban environment. Noise interferes with speech, sleep, learning and work performance and prolonged exposure can induce a hearing loss. Many studies, e.g. Wilson (1963) have identified traffic noise as one of the major sources of annoyance to people whether they are outdoors, at work, or in their home. Efforts to abate traffic noise fall into two main categories: reduction of noise at the source and reduction of the area over which the noise can be regarded as intrusive. Amongst the control measures the placement of barriers between the source and receiver represents a useful technique for modifying noise levels during transmission. Though artificial barriers are most often employed there has been considerable interest in the use of vegetated barriers to act as acoustic buffer zones (see Zulfacar and Clark, 1974).

A useful review of the interaction of noise and plants was given in Robinette (1972) from which it is evident that amount of foliage and the frequency of the sound are amongst the most significant factors affecting noise attenuation. The reduction in sound levels is relatively small except over wide belts of trees, but nevertheless foliage absorbs sound more effectively at high frequencies (to which the ear is more sensitive) so that at moderate sound levels the reduction may be subjectively significant (Moore, 1966).

5.6.1 Noise level measurement
The sound pressure registering on the human ear can vary widely ranging from 2×10^{-5} N m^{-2} to 0.1 N m^{-2}. If p is the acoustic pressure of a given sound and p_0 the reference sound pressure level

(arbitrarily taken as 2×10^{-5} N m^{-2}, i.e. 0 dB) the decibel pressure difference can be expressed as $20 \log_{10} p/p_0$. An increase of 10 dB would be judged by most people as 'about twice as loud' and an increase of about 1 dB is about the minimum perceptible difference in loudness (see Moore, 1966). Most traffic noise studies choose the dB(A) scale (the sound level in decibels as measured on the A scale of a standard sound-level meter) for the physical measure of sound. The selected scale amplifies the frequency components of the sound so that they correspond approximately to the varying sensitivity of the ear at different frequencies; the A scale is applicable for sounds below 55 dB and the B scale for sounds between 55 and 85 dB.

5.6.2 Road traffic noise

Noise levels close to roads carrying large volumes of traffic are likely to be of the order of 75 dB(A). Indeed Road Research Laboratory (1970) reports about 14.5 million people in the United Kingdom living in dwellings exposed to noise levels higher than 65 to 70 dB(A). To give some idea of what this means, a study reported in Wilson (1963) indicated that for exposure levels higher than 55 dB(A) mean energy per 24 hours, the number of individuals considerably disturbed often exceeds 20%. Wilson (1963) suggested that the maximum noise levels inside dwellings should be around 45 dB(A) by day and 35 dB(A) by night and that these values should not be exceeded for more than 10% of the time. In Scandinavia the equivalent noise standards are more stringent and have been recommended as 35 dB(A) (see Organisation for Economic Cooperation and Development, 1971). In most buildings, windows and doors provide the weak path for the air-borne transmission of noise. However, reductions of around 15 to 20 dB(A) are usually achieved by fixed glazing (Scholes and Sargent, 1971). This implies that levels immediately in front of a dwelling should be at the most 50 to 60 dB(A). Comparing these levels with those occurring near the road indicates that one is seeking reductions of up to and around 20 dB(A) during its transmission from the roadside.

One way of achieving reductions of this order is to ensure that dwellings are at a sufficient distance from the road. Figure 5.16 shows how noise levels change with distance relative to the magnitude at a position 30 m from the edge of a nearside road lane. Thus to achieve a reduction of 15 dB(A), dwellings should be no closer than 400 m. A useful representation of the trend shown in fig. 5.16 is the equation:

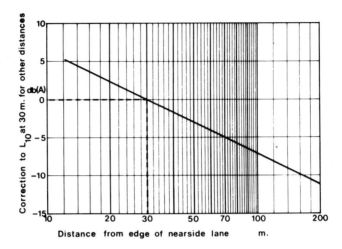

Fig. 5.16 Changes of noise exposure (L_{10}) with distance relative to a reference noise level ($L_{10} \sim 70$ dB(A)) at a position 1 m from a facade at 30 m from the edge of the nearside traffic lane, under wind conditions favourable to the propagation of noise

$$dB(A) = -13.6 \log x + 20.1 \quad (x \text{ in m}) \qquad [5.21]$$

It may not always be feasible to rely on distance alone for achieving the necessary attenuation and in these instances barriers can be valuable. Scholes and Sargent (1971) indicated that the screening of a dwelling from a road can reduce the noise exposure by as much as 20 dB(A), but a more typical reduction is 10 dB(A). Though solid barriers are most often used, plantations of trees represent an alternative. The review by Zulfacar and Clark (1974) showed that attenuation rates with vegetative barriers vary considerably ranging from 1.5 dB per 100 m to 30 dB per 100 m. As a guide O.E.C.D. (1971, page 113) quotes the muffling effect of thick leafy vegetation as 2-3 dB(A) per 100 m and 5-10 dB(A) per 100 m in thick plantations of coniferous trees. Thus it is seen that narrow belts of trees and shrubs are virtually useless as noise attenuators apart from perhaps some psychological advantage in ameliorating the 'effects' of noise. However despite their low attenuation capacity, the planting of trees in the buffer zones adjacent to roads with heavy traffic can bring about a reduction in the minimum distance required to lower the noise to acceptable levels around dwellings. Figure 5.17 shows the correspondence between the minimum distances without obstructions, and the equivalent when the buffer zone is fully planted with trees possessing an attenuation rate of 3 dB(A) per 100 m. If it is assumed that the attenuation arising from the trees and due to the unobstructed distance are additive, the overall noise reduction (when planting over the full distance) can be roughly estimated by the relationship.

$$dB(A) = -\frac{nx}{100} - 13.6 \log x + 20.1 \qquad [5.22]$$

where n is the number of decibels of sound reduction per 100 m of planting. The relationship between the minimum distance x (representing the horizontal axis of fig. 5.17) and the reduced minimum distance, x' (corresponding to the vertical axis of fig. 5.17), can be obtained by eliminating the term dB(A) from equations [5.21] and [5.22] and yields:

$$x = x'\, 10^{\,nx'/1360} \qquad [5.23]$$

Substituting $n = 3$ into equation [5.23] more or less reproduces the trend shown in fig. 5.17. For example, when $x = 400$ m (corresponding to a reduction of 15 dB(A)) the equivalent distance with a fully planted belt is $x' = 170$ m. The principles implicit in equation [5.22] may be applied to other planting patterns.

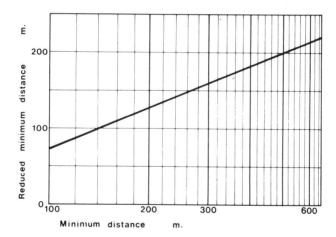

Fig. 5.17 Reduction of minimum distance when planting over the full width with a noise attenuation rate of 3 dB(A) per 100 m

5.7 Shelterbelts and air pollution reduction

The protruding surfaces of vegetation provide a major filtration and reaction interface with the atmosphere, enabling pollutants to be transferred from the air mass into the biosphere. Although the uptake by individual shoots may seem insignificant, their integrated effect over large regions plays an important role in maintaining the cleanliness of the atmosphere.

The character of pollutants is highly variable, but can be divided into gaseous fractions and particulates — the latter embracing both solids and liquids. The gaseous pollutants of major importance are HF, SO_2, NO_2, O_3, Cl_2 and to a lesser extent P.A.N. (peroxyacetyl nitrate) and data suggests that vegetation plays a vital role in their removal (Bennett and Hill, 1975). Undoubtedly, many others such as HCL could be added to the list. Two important pollutants which are known not to be taken up effectively by plants are CO and NO. Uptake of metabolisable pollutants, such as some of the sulphur and nitrogen oxides, is linked to plant metabolism rates and is also influenced by a number of aerodynamic factors related to the wind climate and foliage structure. Hill (1971) demonstrated that vegetation is particularly effective in removing HF, though this may not always be desirable because the accumulation of fluoride in plants may cause injury to sensitive species and poses the threat of causing fluorosis in grazing animals that consume forage over prolonged periods. Beside removing gaseous material, vegetation is effective in trapping particulates, especially those in the larger size ranges. Soil particles, process dust, industrial combustion products and marine salt particles are typically between 1 μm and 10 μm in diameter. Particles in the 0.1-1 μm range frequently represent gases that have condensed to form non-volatile products (Willeke and Whitby, 1975). Particles larger than 10 μm frequently result from mechanical processes such as wind erosion or grinding, or they may be associated with fog water or crop spraying. Many pollens and spores also fall into this category.

In recent years there have been various attempts to exploit the protective value of vegetation for minimising the pollution associated with industrialisation and urbanisation. Zulfacar (1975) points to many examples in which the presence of shelterbelts or green zones has led to significant reductions in the concentration of air-borne contaminants. In parts of Russia the use of vegetative sanitary buffer zones for separating potentially contaminating industrial areas from residential areas has developed as a means of pollution control. The first attempts to find the effectiveness of various widths of the vegetated sanitary buffers were apparently made by Yakovenko who took climatic factors as well as the height of emissions into consideration (Zulfacar, 1975). Goldberg (1948) attempted to categorise the optimal width of buffer zones according to the toxicological potency of the emissions, but the approach was regarded as somewhat arbitrary.

In some regions of the world, forests have been exploited as a means of fog reduction. A notable example is the eastern coast of Hokkaido Island, Japan, which is very often shrouded in fogs in spring and winter; these are the so-called advection fogs coming in from the Pacific Ocean. In the hope of protecting this region against fogs, a vast timbered area has been preserved along the coast. However with increasing population, these forests have been under considerable threat for conversion into agricultural land. As a result, a major research effort (reported in Hori, 1953) was conducted to attempt to arrange the fog-preventing forests in such a manner that they would operate efficiently, yet provide additional space for cultivation over as large an area as possible.

While these case-book examples make interesting reading they are site specific and often lack the necessary information which will allow a particular design to be recast into another situation. Although the art of predicting the pollutant pathways is well advanced, there has been little study of the interaction between pollutants and shelterbelts. This is not surprising when one considers the complexity of a shelterbelt as an aerodynamic structure. Nevertheless it is possible to map out some of the important factors which should be considered when designing vegetated buffer zones. A wider discussion of the use of buffer zones as an air pollution control technique is given in Hagevik *et al.* (1974).

5.7.1 *Specification of deposition*

Many processes contribute to the deposition of air-borne material, diffusion being one of the most important, and for particulates one may also include the effects of inertia and gravity. Often it is difficult to isolate the particular controlling mechanisms especially when physiological features are involved. One way of circumventing this problem is by making use of the concept of deposition velocity defined by:

$$v_d = \frac{\text{flux per unit area }(F)}{\text{air-borne concentration }(C)} \qquad [5.24]$$

The term v_d is characteristic of the transport behaviour in the vicinity of a surface and reflects the ability of the surface to trap air-borne material. The advantage of this formulation is that if v_d can be specified for a particular pollutant-trapping system, then it auto-matically scales the flux to the surface after multiplication by the air-borne concentration. If $v_d = 0$, it implies that there is no transfer of the pollutant to the surface. The concept of deposition velocity is applicable to both gaseous and particulate matter. However when analysing gaseous transfer, analysis is usually based on a resistance analogue (reviewed in Unsworth *et al.* 1976) in which the flux of a gas is envisaged as being driven by a 'potential difference' (difference in gas concentration) and limited by a resistance. By analogy with Ohm's law this transfer is specified by:

$$\text{flux} = \frac{\text{potential difference}}{\text{resistance}} \qquad [5.25]$$

An example of this approach (pertinent to section 2.2.4) is the specification of the flux E of water vapour (i.e. transpiration rate) from a leaf via the stomata. The potential difference is equivalent to the difference between the water vapour concentration e_s in the stomatal cavity (see fig. 2.3) and the water vapour concentration e in the atmosphere. The limiting resistance r is the sum of a component r_a describing the properties of air flow around the leaf and a com-ponent r_{st} related to the dimensions of the stomatal pore. Under these conditions equation [5.25] becomes

$$E \propto \frac{e_s - e}{r} \qquad [5.26]$$

The advantage of the resistance analogue is that it allows a distinc-tion to be made between resistances which are functions of the aero-dynamic properties of a gas exchange system (e.g. the term r_a in equation [2.13]) and resistances which describe physiological or surface properties of plants (such as the term r_s in equation [2.13]). Further discussion of these approaches may be found in the reviews of Chamberlain (1975a, 1975b).

5.7.2 *Pollution dispersion*
A second major item determining the potential design of vegetated buffer zones, concerns pollution dispersion and its interaction with deposition processes. The particular approach to design depends

largely on whether the protected area is close to the source of pollution, or further away. In the former case, one is mainly concerned with the rate of change of air-borne concentration with horizontal distance. Whereas in the latter, pollution levels will be relatively uniform over large areas and it is likely that one will be more concerned with attempts to deplete concentration levels in the lower levels of the atmosphere through the scavenging effects of vegetation. However discussion will be confined to short-range phenomena since it is in this context that buffer zones have been most often considered. An important preliminary is to grasp the rudiments of the atmospheric dispersion of pollutants which is discussed in many basic texts, e.g. Oke (1978). Anybody who has paused to watch smoke emitting from a chimney may appreciate that it is dispersed rapidly due to the motions of the atmosphere. Dispersion takes place both laterally and vertically, the latter generally being the dominant diluting process.

The air-borne concentration of material released continuously from a point source, such as a factory chimney, and unaffected by retention at the ground is commonly represented by the Gaussian dispersion equation:

$$c(x,y,z) = \frac{Q}{2\pi u_h \sigma_y \sigma_z} \exp\left(\frac{-y^2}{2\sigma_y^2}\right)\left[\exp\left(\frac{(z-h)^2}{2\sigma_y^2}\right) + \exp\left(\frac{-(z+h)^2}{2\sigma_z^2}\right)\right]$$

[5.27]

where

$c(x,y,z)$ = concentration at the position (x,y,z) with x,y,z as the downwind, lateral, and vertical distances measured from the source at $(0,0,0)$

Q = source strength (amount of material emitted per unit time)

u_h = wind speed at the source height

h = effective height of release

σ_y, σ_z = the lateral and vertical dispersion coefficients

The dispersion coefficients are often represented by empirical equations of the form $\sigma = ax^b$ where a and b are constants depending on wind and temperature effects in the lower atmosphere. Their importance is that they scale the degree of spread. For most purposes the lateral spread can be adequately specified by $\sigma_y = 0.1\ x$ which indicates that the plume spreads linearly with distance. However the vertical spread is sensitive to turbulent mixing in the atmosphere. It is least in calm conditions, typical of dawn or dusk (termed stable conditions) and is most in conditions of strong sunshine and little wind (termed unstable conditions). For intermediate situations, usually associated with overcast conditions and moderate to strong winds, the lower atmosphere is well-mixed and temperature effects are negligible (termed adiabatic or neutral conditions). Data shown in table 5.2 gives an indication of the dispersive capability of the atmosphere for different classes of stability (see Pasquill, 1974).

Table 5.2 Coefficients of the empirical dispersion equation $\sigma_z = a\,x^b$ for estimating vertical diffusion near the ground over terrain of moderate roughness ($z_0 = 0.5$ m) as a function of the Pasquill stability category and applicable for distances up to 2 km (based on data from Pasquill (1974) with σ_z and x in metres)

Stability category	a	b	Remarks
A very unstable	0.34	0.91	low wind speed and strong insolation
B unstable	0.29	0.85	
C slightly unstable	0.33	0.77	overcast
D neutral	0.25	0.75	overcast, strong winds
E slightly stable	0.21	0.71	
F very stable	0.15	0.67	at night-time with low wind speed and little cloud

5.7.3 Broad buffer zones

Having briefly reviewed dispersion and deposition processes we can now consider their combined effect in so far as they determine the ground level concentration at a point (x) located on the downwind edge of a broad absorbing buffer zone. By considering the fractional reduction of air-borne concentration with distance, it is possible to make deductions about the necessary width of a buffer zone in order to reduce the air-borne concentration to a desired level.

To simplify the discussion, we can consider the case of a ground source (i.e. $h = 0$) and ignore the lateral spreading by integration over the y direction, which yields:

$$\langle c(x,y) \rangle = \frac{Q_0}{\sigma_z u} \left[\frac{2}{\pi} \right]^{1/2} \exp\left[\frac{-z^2}{2\sigma_z^2} \right] \qquad [5.28]$$

where $<c(x,z)>$ is the cross-wind integrated concentrations. Equation [5.28] also represents the concentration distribution which would be expected downwind of a line source, e.g. the pollution emitted from traffic on a busy road.

When surface deposition occurs the amount of material in the air is slowly depleted, and may be specified by:

$$Q(x) = Q_0 \exp \left[\frac{-v_d}{u_h} \left(\frac{2}{\pi}\right)^{1/2} \frac{1}{1-b} \left(\frac{x}{\sigma_z(x)} - \frac{x_1}{\sigma_z(x_1)}\right) \right] \quad x > x_1$$

$$= Q_0 \text{ for } x \leqslant x_1 \qquad\qquad\qquad\qquad\qquad [5.29]$$

The above expression is a modified form of equation [5.31a] from Pasquill (1974); it assumes that deposition does not occur until the plume reaches the position x_1 which is considered as the upwind edge of an absorbing buffer region. By replacing the term Q_0 in equation [5.28] by $Q(x)$ as defined above, the modified equation [5.28] yields $<c(x,z)>_{dep}$ when surface depletion is included. The effect of the absorbing region can be demonstrated by evaluating the ratio:

$$\frac{<c(x,0)>_{dep}}{<c(x,0)>} = \exp \left[\frac{-v_d}{u_h} \left(\frac{2}{\pi}\right)^{1/2} \frac{1}{1-b} \left(\frac{x}{\sigma_z(x)} - \frac{x_1}{\sigma_z(x_1)}\right) \right]$$

$$\text{for } x > x_1 \qquad\qquad\qquad [5.30]$$

which shows the ground concentration (including deposition) as a ratio of its equivalent level with deposition absent, with $x - x_1$ as the width of the buffer zone. Data shown in table 5.3 illustrates the effect of shelterbelt width on weakly and strongly absorbed pollutants in adiabatic conditions. It is seen that the buffer region is only really effective for strongly absorbed pollutants.

When inspecting claims of the apparent value of shelterbelts in reducing pollution levels in the vicinity of its source, it is extremely important to distinguish between the reductions 'caused by deposition' and those generated by turbulent dispersion. In the absence of deposition, the fractional reduction due to dispersion alone is obtained from equation [5.28]:

$$\frac{<c(x_1,0)>}{<c(x_1,0)>} = \frac{\sigma_z(x_1)}{\sigma_z(x)} = \left(\frac{x_1}{x}\right)^b \qquad\qquad [5.31]$$

which compares the concentration levels occurring at distances x and x_1. For example, the data shown in table 5.2 for neutral conditions yields $<c(700,0)>/<c(400,0)> = 0.66$ and $<c(1000,0)>/<c(400,0)> = 0.50$.

In both cases the dilution attributable to distance is greater than the reduction caused by deposition (see table 5.3). If the same analysis is carried out, but including the effect of deposition then the ratio of the concentrations at the distances x and x_1 is given by:

$$\frac{<c(x,0)>_{dep}}{<c(x_1,0)>} = \frac{<c(x,0)>_{dep}}{<c(x,0)>} \quad \times \quad \frac{<c(x,0)>}{<c(x_1,0)>} \qquad [5.32]$$

equation [5.30]   equation [5.31]

Table 5.3   Fractional reduction in air-borne concentration as a function of barrier width and absorption characteristics

Deposition velocity (m s^{-1})	barrier width	$<c(x,0)>_{dep}/<c(x,0)>$ 300 m	600 m
0.01		0.98	0.96
0.08		0.84	0.74

Data was calculated using equation [5.30] with $x_1 = 400$ and $x = 700, 1000$ m with $v_d = 0.01. 0.08$ m s^{-1}, $u_h = 4$ m s^{-1} and σ_z based on table 5.2 for neutral conditions in fairly rough terrain, perhaps representative of the urban-shelterbelt transition region ($z_0 = 0.5$ m)

5.7.4 Narrow barriers

When considering the pollution reduction by a narrow barrier the first stage is to evaluate the mass balance across a representative area, A, perpendicular to the wind direction. From equation [5.24] it is evident that the deposited flux depends on the effective trapping area encountered by the wind during its path through the barrier. This can be specified by the expression:

$$\text{deposited flux} = v_d \bar{c} f A \qquad [5.33]$$

in which v_d is the local deposition rate, $\bar{c}$ the mean local concentration and f a proportionality coefficient representing the effective trapping area per unit face area. For order of magnitude purposes, $\bar{c} = (c + c')/2$ where c is the incoming concentration and c' the reduced concentration immediately behind the barrier. With these factors the mass balance reads:

$$\underset{\text{incoming}}{\overline{u}\, c\, A} \quad - \quad \underset{\text{deposited}}{v_{\text{d}}\, \overline{c}\, f\, A} \quad = \quad \underset{\text{outgoing}}{\overline{u}\, c'\, A} \tag{5.34}$$

in which $\overline{u}$ is the local wind speed within the barrier (for a belt of medium permeability this may be taken as roughly half the open field wind speed (see fig. 5.2)). Transformation of equation [5.34] yields the concentration ratio:

$$\frac{c'}{\overline{c}} = \frac{1 - f\, v_{\text{d}}/2\overline{u}}{1 + f\, v_{\text{d}}/2\overline{u}} \tag{5.35}$$

and shows its explicit dependence on the parameters f, v_{d} and $\overline{u}$.

Although there may be a reduction of concentration immediately behind the barrier the filtered air rapidly mixes with air passing over the barrier and restores the concentration to its original value. At present it is not possible to predict the behaviour within this regime except by using complex modelling techniques. Nevertheless it is reasonable to suppose that the protected distance will be similar to that arising with wind speed. Between the barrier and the limit of protection, the air-borne concentration must increase.

5.7.5 Perspective

In the preceding section it was evident that simple models of pollution transfer enable us to gauge the likely influence of a shelterbelt or stands of vegetation on pollution levels downwind of a source. The particular situations examined illustrate some of the basic principles which must be taken into consideration in pollution control and require further development. Nevertheless they permit order-of-magnitude calculations for 'worst case' situations in which the pollution plume blows directly to the recipient.

Of the parameters involved, most can be measured directly or estimated from the wealth of literature on pollution transfer. In the case of the deposition velocity, care should be taken to distinguish between its local value characteristic of a particular trapping element, and its value as a bulk parameter representative of the trapping ability of a complete stand of vegetation. The relationship between the two was analysed in Bache (1979). Extensive reviews by McMahon and Denison (1979) and Sehmel (1980) give guidance to the deposition rates of different gases and categories of particulate matter over a variety of capture surfaces.

As a trapping medium, foliage offers greater potential for the trapping of larger particulate matter (say > 5 μm in diameter) than for gases whose deposition velocity is limited (in practice) to about 0.05 m s^{-1}. However there is considerable variability. For particulate matter the deposition rate is critically dependent on the wind speed and particle diameter (Belot and Gauthier, 1975). The latter study is of particular interest because it demonstrated the superior trapping ability of a surface with a fibrous structure (as in pine shoots) compared with a bluff-shaped structure typical of broad-leaf species.

Finally, if plants are to be treated as a means of reducing air pollution then it is vital to consider the effect of the pollutants on the plant. Observations of injury on sensitive plant species have provided a valuable means of monitoring pollutant emissions from a source, since the symptoms which usually develop are characteristic of the specific pollutant (Jacobson and Hill, 1970). The latter survey is extremely useful, because it examines the sensitivity of a wide range of plants to the major air pollutants and facilitates the selection of suitable species for planting in a polluted environment. Pollutants generally pose the greatest risk to fruit trees and many of the economically important conifers, whereas agronomic and ornamental plants are for the most part in less danger from chronic pollutant effects (Guderian, 1977).

Example 5.3 Pollution reduction by narrow barriers

It is proposed to construct a narrow, but fairly dense shelterbelt using fast-growing conifers to reduce the aerosol drift arising from the spray application of waste-water on a land application site. Preliminary analysis indicates that at the proposed siting of the barrier the aerosols are in droplet form with diameters in the range 5-30 μm. Estimate the potential reduction in the aerosol concentration assuming that the shelterbelt is sufficiently high to intercept the spray plume. The average wind speed for the area is 3 m s^{-1}.

When considering the trapping of particulate material it is useful to transform equation [5.24] using the substitution $E = v_d/u$ which defines the trapping coefficient (see Chamberlain, 1975b):

$$\frac{c'}{c} = \frac{1 - fE/2}{1 + fE/2} \qquad [5.36]$$

Data from many studies, e.g. May and Clifford (1967) and Belot and Gauthier (1975), show that the impaction parameter depends largely on S, an inertia parameter defined by:

$$S = \frac{\rho_p u d_p^2}{18 \nu \rho_a \lambda} \tag{5.37}$$

in which ρ_p, ρ_a are the density of the particle and air respectively, d_p is the particle diameter, ν the kinematic viscosity of air, λ a characteristic dimension of the trapping element and u the local wind speed. For situations in which $S > 0.1$, E can be roughly estimated using the relationship:

$$E = \frac{S^2}{(S+0.4)^2} \tag{5.38}$$

which is a rough fit of data shown in May and Clifford (1967) for impaction on cylinders and ribbons. If it is assumed that $\rho_p = 1000$ kg m^{-3}, $\rho_a = 1.24$ kg m^{-3}, $u = 1.5$ m s^{-1} (taken as half to wind speed in the open) $d_p = 5$ and 30 μm, $\nu = 1.42 \times 10^{-5}$ m^2 s^{-1} and $\lambda = 2$ mm (see Belot and Gauthier, 1975) then S may be calculated using equation [5.37] leading to the values shown in column (2) of table 5.4. Corresponding estimates of the impaction parameter based on equation [5.38] are shown in column (3) and the final column is determined by equation [5.36] assuming $f = 0.8$ to be representative of a fairly dense barrier. The analysis suggests that the belt has some merit in reducing concentration levels provided the droplets are sufficiently large, but offers no protection against aerosols with diameters < 5 μm.

Table 5.4   Reduction of concentration levels with narrow barriers

d_p (μm) (1)	S (2)	E (3)	c'/c (4)
5	0.059	0.017	0.99
30	2.13	0.77	0.53

5.8 Design strategy

From the previous discussion it is apparent that there are many avenues which can be explored to elucidate the role of shelterbelts and windbreaks. The following sections seek to provide general insight into the design strategy for particular situations.

5.8.1 Shelterbelts and buildings

If a shelterbelt is located at roughly $3H$ to $6H$ from a building this will tend to take best advantage of its capacity as a windbreak. However it is also necessary to consider whether a belt located within this range is going to cause excessive shading. Some guide of shadow lengths on level ground can be obtained by evaluating equation [5.20] for particular times of the day and during the year. A useful case-book example is described in Caborn (1965) which, for U.K. conditions, suggests that plantings should be no closer than about $4H$ except on the north-facing side where shading is less of a problem. The planting of trees too close to buildings may also cause structural damage by generating excess turbulence or physical contact, particularly in the foundations. A number of windbreaks situated around the building are useful for cases of variable wind direction but care should be taken to avoid openings which promote the funnelling of wind. For a more detailed review of the interrelationship of planting patterns and airflow around buildings the reader is referred to Robinette (1972).

An additional benefit of locating shelterbelts near buildings is their potential value in reducing the rate of heat loss, thus saving fuel; this arises because reductions in wind speed diminish the heat transmission coefficient, which is sensitive to wind speed. This aspect of shelterbelts has gained considerable attention in the glasshouse industry. For example, in the United Kingdom glasshouses use about 35% of the fuel used directly in agriculture; also, heating accounts for 30-35% of the cost of production of many crops. Considerable reduction in heat loss can be effected by the provision of efficient shelter; the potential benefits are evaluated in Sheard (1975).

5.8.2 Shelterbelts for farmland

When planting for shelter on a farm, one of the first requirements is to assess the particular needs of the farm, whether dealing with livestock, crops or both. As a general rule plantings should blend in with the existing landscape and it is useful to take advantage of existing hedgerows and tree belts as part of the overall pattern for increasing shelter. Wasteful use of land can be avoided by making use of existing boundaries such as those delineating fields or changes of ownership or perhaps bordering roads and streams. Particular attention should be paid to the local climate and to the general topography of the area. It should be stressed that damage following a particular wind may be due to other factors accompanying it, e.g. low temperatures, and such conditions are not necessarily associated with the prevailing wind direction; this may influence the eventual siting of the belt.

In hilly country, belts should be parallel or perpendicular to the contours (van Eimern *et al.* 1964). Planting on the contour fits in with existing field cropping techniques of allowing runoff to penetrate, reducing soil erosion. However an adverse effect of shelterbelts planted across the slope is that they impede air drainage on cold nights and so increase frost risks (see fig. 5.18). These problems can be reduced by keeping the base of shelterbelts fairly open to allow the cold air to escape.

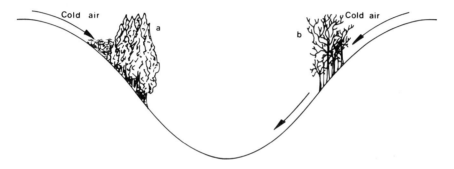

Fig. 5.18 (a) Dense windbreaks create frost pockets; (b) penetrable windbreaks allow cold air to drain away and are important at the foot of sloping gardens and orchards

In hilly country, winds tend to follow the direction of the valleys, and in undulating country with gentle slopes (say < 1:5) winds tend to follow the contours. In both situations belts are best sited so as to run at right angles to the wind, i.e. up and down the slopes. Strongest winds are likely to occur on the windward slope and on the crests of hills. For gently undulating country, shelterbelts are best sited on the crest of folds and ridges or just on the leeward slope of a hill. However for country with a sharper relief, the best guidance is to plant just to the leeward of a ridge or eminence and to avoid both the crest and foot of the slope of a hill (Ministry of Agriculture, Fisheries and Food, 1977).

5.8.3 Shelterbelts and roads
In addition to making a road look attractive, trees and shrubs have a number of functional uses. Ground cover is of course necessary to reduce wind and water erosion. Plants may also serve as indicators to accentuate the position of road structures and junctions. However our main concern here is with their use as barriers or screens. Previous

discussion has focused on their role in noise abatement. Here, we consider their value in the control of glare, wind and snow, drawing attention to the safety aspects of roadside planting and of their potential in causing road damage. For a wider discussion of the landscaping of roads the reader is referred to texts such as Crowe (1960), British Road Federation (1963), Robinette (1972) and Hackett (1979).

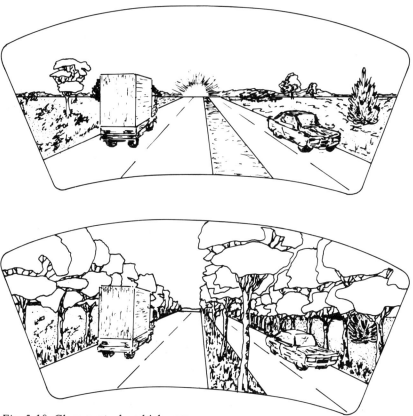

Fig. 5.19 Glare control on highways

GLARE CONTROL

Roadside plantings, when in leaf, are a useful means of reducing the glare from bright lights, which disturb road users or the inhabitants in the surrounding area. At night street lights and vehicle headlights produce glare and reflections. Dazzle from the early morning and late afternoon sun poses a particular safety hazard especially when the roads are wet. Strategic planting of the type shown in fig. 5.19 may help to obviate such problems and is discussed further in Robinette (1972).

Planting in the central reservation of a dual carriageway road can overcome the dazzle problem from oncoming headlights. Trees should not be planted unless the central strip is sufficiently wide. Further, if the width is too small, the plantings may suffer from the effects of salting against ice (see Westing, 1969). Hackett (1979) suggests that plantings should be kept at least 5 m from the edge of the carriageway to avoid damage. When the road curves, or there is a gradient, planting should be continuous and at least 1.8 m in height. Otherwise gaps may be left provided there are no wind problems.

WINDBREAKS

Wind conditions on roads rarely call for control except around the ends of cuttings and embankments which are areas prone to excessive gusting on windy days. Caborn (1965) suggests that these conditions can be improved by planting elongated clumps of trees in accord with the pattern in fig. 5.20. These should aim to filter the wind rather than deflect it and ideally should be thicker next to the end of the formation and thinner further away.

SAFETY AND DAMAGE FACTORS

Highway engineers generally do not like trees next to roads, particularly high-speed routes because of the safety hazard and the potential damage caused by root-spread.

One particular problem is the cracking and settlement of the road surface promoted by soil drying arising from root suction and transpiration. In the aftermath of a prolonged drought in 1946 and 1947 in southern England, a survey was carried out to assess the road damage caused by vegetation. Results were reported in Croney and Lewis (1949) who made a number of tentative recommendations. They suggested that fast-growing trees should not be used within 15 m of the carriageway on heavy clay soils and only small-growing trees and shrubs should be used within 3 m. However over loam and sandy soils, trees and shrubs of all suitable types can be used freely provided that the expanding trunk and root formation does not cause deformation to the pavement or kerb.

Little research has been carried out on the safety aspects of trees alongside roads though it is generally agreed that shrubs and small trees are preferable if they are to be treated as crash barriers. For shrubs, some useful data was collated in White (1953) following a series of crash tests to evaluate the ability of plantings of *Rosa multiflora* to absorb impact energy.

Another important aspect of safety concerns the maintenance of visibility and it is generally desirable to maintain a clear area outside

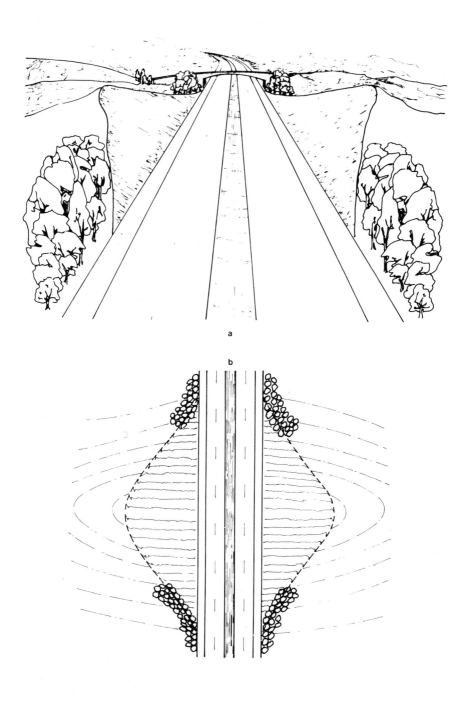

a

b

Fig. 5.20 Suggested arrangement of planting to check cross-wind gusts near cuttings on highways: (a) general view; (b) plan of arrangement

the carriageway before planting of at least 4 m for high-speed roads and just over 2 m for slow-speed roads (Hackett, 1979).

SNOW CONTROL

Shelterbelts are exploited for keeping roads and other routes of communication open and increasing use is being made of various snow-collecting fences. The designs and performances are similar, but trees have a considerable height advantage and this enables them to control larger volumes of snow over greater distances. Though expensive to install, the costs of snow fences may be cheap compared with the costs of snow clearance. Snow fences have a number of advantages in that they take up little land space, are immediately effective, standardised in behaviour, and flexible in their use.

The general philosophy of protecting roads is to induce a drift on the windward side of the road ensuring that the distance between the windbreak and the road is greater than the expected drift length.

According to German research (Bekker, 1947) the drift length L(m) is related to the fence height H(m) by the equation:

$$L = \frac{11 + 1.5H}{k} \qquad\qquad [5.39]$$

where k is a function of the fence density (relative fraction of area covered by slats in a plane perpendicular to the wind direction) being unity for a density of 50%; 1.28 for a density of 70%; and $k < 1$ (e.g. 0.8) for a 35% barrier (Caborn, 1965). A further 5 m should be allowed when estimating L, to take account of variability. It is generally agreed that the optimum density for a snow barrier is about 50%. In Britain, Caborn (1967) remarks that fences are often set about 30 m back from the road, but in other climates a greater distance is sometimes preferred (see van Eimern et al. 1964).

Hedges represent an alternative to fences and may be cheaper in the longer term. For tallish hedges fairly open to the wind, Caborn (1967) states that drifting rarely extends more than 18 m, and the bulk of the snow invariably lodges within 12 m on the leeward side. Shelterbelts with an underwood normally collect snow within 30 m on the lee side (and usually over smaller distances). Caborn also emphasises the importance of including a light row of shrubs to sieve the snow and initiate a drift. Rows of trees right beside the road keep snow away provided the space between the trunks is free of branches and bushes (Kuhlewind et al. 1955). The penalty of retaining impermeable hedges alongside a road is illustrated in fig. 5.21.

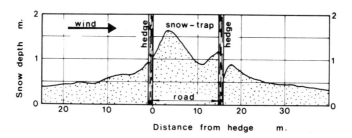

Fig. 5.21 Snow accumulation by two hedges planted too close to a road

5.9 Structural aspects

While previous sections have concentrated on the fundamental aspects of shelterbelt design, it remains a matter of translating theory into practice by selecting the most appropriate tree species and developing the structure of the belt. A full discussion of this aspect is given in Caborn (1965). According to the Ministry of Agriculture, Fisheries and Food (1977) three main aims must be borne in mind:
(a) planting of a main species which will quickly reach the desired height;
(b) planting of other species which will maintain an effective density between the crowns of the main species and the ground;
(c) permanence of the belt as an effective screen.

The ideal of about 40% permeability which satisfies many purposes implies that the belt must be narrow (say up to three rows of trees). In theory, a single row of trees is as effective as several rows when serving as a windbreak, and has the advantage of saving space. However, it is more difficult to maintain the required structure, particularly if some of the trees fail. In this respect multiple-row shelterbelts offer greater flexibility and pose fewer problems. In more exposed situations or where catering for snow storage, more rows may be required, but this depends on the availability of land.

5.9.1 *Pure and mixed plantings*
Though shelterbelts with a single species are easier to design and manage it is advisable to plant more than one species, since their character and density vary at different heights. Deciduous trees are leafless in winter when shelter is generally most needed, and unless they are very twiggy, they will fail to slow wind speed sufficiently. Evergreens tend to be too dense when young and because of their

greater density are more likely to be blown over. According to
Ministry of Agriculture, Fisheries and Food (1977), a mixture of
deciduous and evergreen species usually gives the best results and
will maintain the density of the belts during the winter. However
Caborn (1965, page 152) does not support this view when considering mixtures within single rows and favours a pattern of trees
and shrubs.

Mixed plantings cannot be planned without some knowledge of
the growing characteristics and physical needs of different species.
For example, some species are fast-growing and light-demanding,
whereas others are slow-growing and need to be shaded and sup-
pressed unless they are naturally tolerant to shade.

Permanence requires careful planning as the growth of individual
species varies, but it is useful to aim for a belt consisting of trees of
all sizes and ages which can be cut and replanted as desired. Outside
rows bear the brunt of the exposure and it is crucial that they should
be wind-firm.

5.9.2 Species selection

The best guide in deciding on the species to plant may be obtained
from direct observation of existing vegetation or from comparable
situations. This information can be supplemented by local guides
such as Ministry of Agriculture, Fisheries and Food (1977, 1979,
1980) covering shelterbelt and windbreak design in the United
Kingdom, or van Rensburg (1973) for comparable plantings in South
Africa.

5.10 References

Bache, D.H. (1979) 'Particulate transport within plant canopies. II. Prediction of
    deposition velocities'. *Atmospheric Environment* **13**, 1681-87.
Bates, C.G. (1944) 'The windbreak as a farm asset'. *Farmers' Bulletin*. U.S.
    Department of Agriculture 1405.
Belot, Y. and Gauthier, D. (1975) 'Transport of micronic particles from the
    atmosphere to foliar surfaces'. In de Vries, D.A. and Afgan, N.H. *eds. Heat
    and Mass Transfer to the Biosphere. Part 1. Heat Transfer in the Plant
    Environment*. Washington: Scripta, 583-91.
Bekker, M.G. (1947) *Snow Studies in Germany*. Canada: Department of National
    Defence (Army), 53-63.
Bennett, J.H. and Hill, A.C. (1975) 'Interactions of air pollutants with canopies
    of vegetation'. In Mudd, J.B. and Kozlowski, T.T. *eds. Responses of Plants to
    Air Pollution*. New York: Academic Press.

Black, A.L. and Siddoway, F.H. (1971) 'Tall wheatgrass barriers for soil erosion control and water conservation'. *Journal of Soil and Water Conservation* **26**(3), 107-11.

Black, A.L. and Siddoway, F.H. (1975) 'Snow trapping and crop management with tall wheatgrass barriers in Montana'. *Snow Management in the Great Plains*. Publication No. 73, University of Nebraska, Lincoln Agricultural Experiment Station: Research Committee Great Plains Agricultural Council, 128-37.

British Road Federation (1963) *Landscaping of Motorways*. London: British Road Federation.

Businger, J.A. (1975) 'Aerodynamics of vegetated surfaces'. In de Vries, D.A. and Afgan, N.H. eds. *Heat and Mass Transfer in the Biosphere. Part 1. Heat Transfer in the Plant Environment*. Washington: Scripta, 139-65.

Caborn, J.M. (1957) *Shelterbelts and Microclimate*. Forestry Bulletin No. 29. Edinburgh: Her Majesty's Stationery Office.

Caborn, J.M. (1965) *Shelterbelts and Windbreaks*. London: Faber and Faber.

Chamberlain, A.C. (1975a) 'Pollution in plant canopies'. In de Vries, D.A. and Afgan, N.A. eds. *Heat and Mass Transfer in the Biosphere. Part 1. Transfer Processes in the Plant Environment*. Washington: Scripta, 561-82.

Chamberlain, A.C. (1975b) 'The movement of particles in plant communities'. In Monteith, J.L. ed. *Vegetation and the Atmosphere*. Volume 1. New York: Academic Press, 155-203.

Chepil, W.S. and Woodruff, N.P. (1963) 'The physics of erosion and its control'. *Advances in Agronomy* **15**, 211-302.

Counihan, J. (1975) 'Adiabatic atmospheric boundary layers: a review and analysis of existing data from the period 1880-1972'. *Atmospheric Environment* **9**, 871-905.

Croney, D. and Lewis, W.A. (1949) 'The effect of vegetation on the settlement of roads'. *Proceedings of the Conference, Biology in Civil Engineering*. London: Institution of Civil Engineers, 195-222.

Crowe, S. (1960) *The Landscape of Roads*. London: Architectural Press.

Frank, A.B. and George, E.J. (1975) 'Windbreaks for snow management in north Dakota'. *Snow Management of the Great Plains*. Publication No. 73. University of Nebraska, Lincoln Agricultural Experiment Station: Research Committee Great Plains Agricultural Council, 144-54.

George, E.J. (1943) 'Effects of cultivation and number of rows on survival and growth of trees in farm windbreaks on the northern Great Plains'. *Journal of Forestry* **41**, 820-28.

Goldberg, M.S. (1948) 'The sanitary protection of air'. *Gigiena i Sanitariya (MIDZIG)* **119**. Moscow, U.S.S.R., Appendix 3.

Grace, J. (1977) *Plant Response to Wind*. New York: Academic Press.

Great Plains Agricultural Council (1976) *Shelterbelts on the Great Plains*. Publication No. 78. University of Nebraska, Lincoln: Great Plains Agricultural Council.

Guderian, R. (1977) *Air pollution: Phytotoxicity of Acidic Gases and its Signifi-*

cance in *Air Pollution Control*. New York: Springer-Verlag.

Hackett, B. (1979) *Planting Design*. London: Spon.

Hagen, L.J. (1976) 'Windbreak design for optimum erosion control'. *Shelterbelts on the Great Plains*. Publication No. 78. University of Nebraska, Lincoln: Great Plains Agricultural Council, 31-36.

Hagevik, G., Mandelker, D.R. and Brail, R.K. (1974) *Air Quality Management and Land Use Planning: Legal, Administrative and Methodological Perspectives*. New York: Praeger.

Hill, A.C. (1971) 'Vegetation: a sink for atmospheric pollutants'. *Journal of the Air Pollution Control Association* 21, 341-46.

Hori, T. (ed.) (1953) *Studies on Fogs in Relation to Fog-Preventing Forests*. Sapporo, Japan: Tanne Trading Company.

Iqbal, M., Khatry, A.K. and Seguin, B. (1977) 'A study of the roughness effects of multiple windbreaks'. *Boundary Layer Meteorology*, 11, 187-203.

Jacobson, J.S. and Hill, A.C. eds. (1970) *Recognition of Air Pollution Injury to Vegetation. A Pictorial Atlas*. Pittsburg, Pennsylvania: Air Pollution Control Federation.

Kreutz, W. (1958) 'Welches ist der optimale Abstand für Windschutzstreifen' [What is the optimal spacing for shelterbelts?] . *Holzzucht* 12, 10.

Kuhlewind, C., Bringmann, K. and Kaiser, H. (1955) *Richtlinien für Windschutz. I. Teil Agrarmeteor. und landw. Grundlagen* [Directions for Shelterbelts. Part 1. Agrometeorological and Agricultural Basis] . Frankfurt: Verlag Deutsche Landwirtschafts-Gesellschaft.

Lettau, H. (1969) 'Note on aerodynamic roughness parameter estimation on the basis of roughness element description'. *Journal of Applied Meteorology* 8, 828-32.

McMachon, T.A. and Denison, P.J. (1979) 'Empirical atmospheric deposition parameters – a survey'. *Atmospheric Environment* 13, 571-85.

Marshall, J.K. (1967) 'The effects of shelter on the productivity of grassland and field crops'. *Field Crop Abstracts* 20, 1-14.

May, K.R. and Clifford, R. (1967) 'The impaction of aerosol particles on cylinders, spheres, ribbons and discs'. *Annals of Occupational Hygiene* 10, 83-95.

Ministry of Agriculture, Fisheries and Food (1977) *Shelter Belts for Farmland*. F.E.F. Leaflet No. 15. London: Her Majesty's Stationery Office.

Ministry of Agriculture, Fisheries and Food (1979) *Windbreaks*. Agricultural Development Advisory Service, Publication No. HG21. Edinburgh: Her Majesty's Stationery Office.

Ministry of Agriculture, Fisheries and Food (1980) *Windbreaks for Glasshouses and Plastic Structures*. Agricultural Development Advisory Service, Booklet 2284. Pinner, U.K.: M.A.F.F. (Publications).

Moore, D.J. (1975) 'A simple boundary-layer model for predicting time mean ground-level concentrations of material emitted from tall chimneys'. *Proceedings of the Institution of Mechanical Engineers* 189 (4), 33-43.

Moore, J.E. (1966) *Design for Noise Reduction*. London: Architectural Press.

Nägeli, W. (1946) 'Weitere Untersuchungen über die Windverhältnisse im Bereich von Windschutzanlagen' (German with French Summary) [Further investigation on wind conditions in the range of shelterbelts]. *Mitteilungen des Schweizerischen Anstalt für das forstliche Versuchswesen* **24**, 659-737.

Nägeli, W. (1953) 'Untersuchungen über die Windverhältnisse im Bereich von Schfrohrwänden' (German with French summary) [Investigations on the wind conditions in the range of narrow walls of reed]. *Mitteilungen der Schweizerischen Anstalt für das forstliche Versuchswesen* **29** (2), 213-66.

Oke, T.R. (1978) *Boundary Layer Climates*. London: Methuen.

Organisation for Economic Cooperation and Development (1971) *Urban Traffic Noise: Strategy for an Improved Environment*. Report on the Consultative Group on Transportation Research, Paris: O.E.C.D.

Pasquill, F. (1974) *Atmospheric Diffusion*. 2nd edn. Chichester, U.K.: Horwood.

Patten, O.M. (1956) *Shelterbelts on Your Farm*. Montana State College, Bozeman: Extension Service.

Plate, E.J. (1971) 'The aerodynamics of shelter belts'. *Agricultural Meteorology* **8**, 203-07.

Road Research Laboratory (1970) *Report of Working Group on Research into Road Traffic Noise: A Review of Road Traffic Noise*. Crowthorne, U.K.: Road Research Laboratory, Department of the Environment.

Robinette, G.O. (1972) *Plants/People/and Environmental Quality*. Washington, D.C.: U.S. Department of the Interior, National Park Service.

Rosenberg, N.J. (1976) 'Effects of windbreaks on the microclimate, energy balance and water use efficiency of crops growing in the Great Plains'. *Shelterbelts on the Great Plains*. Publication No. 78. University of Nebraska, Lincoln: Great Plains Agricultural Council, 49-56.

Scholes, W.E. and Sargent, J.W. (1971) *Designing against Noise from Road Traffic*. Current Paper 20/71. Watford, U.K.: Building Research Station, Department of the Environment.

Seginer, I. (1975) 'Flow round a windbreak in an oblique wind'. *Boundary Layer Meteorology* **9**, 133-41.

Seguin, B. and Gignoux, N. (1974) 'Etude expérimentale de l'influence d'un réseau de brise-vent sur le profil vertical de vitesse du vent' [An experimental study of wind profile modification by a network of shelterbelts]. *Agricultural Meteorology* **13**, 15-23.

Sehmel, G.A. (1980) 'Particle and gas dry deposition: a review'. *Atmospheric Environment* **14**, 983-1011.

Sheard, G.F. (1975) 'The effects of wind on glasshouse production'. *Proceedings of the Fourth Symposium on Shelter Research, Warwick University*. London: Agricultural Development Advisory Service, Ministry of Agriculture, Fisheries and Food, 23-31.

Skidmore, E.L. (1965) 'Assessing wind erosion forces: directions and relative magnitudes'. *Soil Science Society of America, Proceedings* **29**, 587-90.

Skidmore, E.L. and Hagen, L.J. (1970) 'Evapotranspiration and the aerial environment as influenced by windbreaks'. *Proceedings of Great Plains*

Evapotranspiration Seminar, Bushland, Texas. Publication No. 50. University of Nebraska, Lincoln: Great Plains Agricultural Council, 339-68.

Skidmore, E.L. and Hagen, L.J. (1973) 'Potential evaporation as influenced by barrier-induced microclimate'. *Ecological Studies 4*. New York: Springer-Verlag, 237-44.

Skidmore, E.L. and Hagen, L.J. (1977) 'Reducing wind erosion with barriers'. *Transactions of the American Society of Agricultural Engineers* **20**, 911-15.

Skidmore, E.L. and Woodruff, N.P. (1968) *Wind Erosion Forces in the United States and Their Use in Predicting Soil Loss*. Agricultural Handbook No. 346. Washington, D.C.: Agricultural Research Service, U.S. Department of Agriculture.

Stoeckler, J.H. and Dortignag, E.J. (1941) 'Snowdrifts as a factor in growth and longevity of shelterbelts in the Great Plains'. *Ecology* **22**, 117-24.

Tabler, R.D. (1975) 'Estimating the transport and evaporation of blowing snow'. *Snow Management on the Great Plains*. Publication No. 73. University of Nebraska, Lincoln Agricultural Experiment Station: Research Committee Great Plains Agricultural Council, 85-104.

Tanner, H.C.H. and Nägeli, W. (1947) 'Wetterbeobachtungen und Untersuchungen über die Windverhältnisse im Bereich von Laub – und Nadelholzschutzstreifen' [Weather observations and investigations in the range of shelterbelts and conifers] . *Jahresbericht der Melioration der Rheinebene* (Schweiz) 28-43.

Unsworth, M.H., Biscoe, P.V. and Black, V. (1976) 'Analysis of gas exchange between plants and polluted atmospheres'. In Mansfield, T.A. *ed. Effects of Air Pollutants on Plants*. Cambridge: Cambridge University Press.

van Eimern, J., Karschon, R., Razumova, L.A. and Robertson, G.W. (1964) *Windbreaks and Shelterbelts*. Technical Note No. 59. Geneva: World Meteorological Organisation.

van Rensburg, A.D. (1973) *Shelterbelts and Windbreaks in South Africa*. Pamphlet No. 118. Pretoria: Department of Forestry.

Westing, A.H. (1969) 'Plants and salt in the roadside environment'. *Phytopathology* **59**, 1174-81.

White, A.J. (1953) *Study of Marginal and Median Tree and Shrub Planting As Safety Barriers on Highways*. South Lee, New Hampshire: Motor Vehicle Research Inc.

Willeke, K. and Whitby, K.T. (1975) 'Atmospheric aerosols: size distribution interpretation'. *Journal of the Air Pollution Control Association* **25**, 529-34.

Wilson, A. (1963) *Noise*. London: Her Majesty's Stationery Office.

Woodruff, N.P. and Siddoway, F.H. (1965) 'A wind erosion equation'. *Soil Science of America, Proceedings* **29**, 602-08.

Woodruff, N.P. and Zingg, A.W. (1952) *Wind-tunnel Studies of Fundamental Problems Related to Windbreaks*. Report No. USDA, SCS-CP-112. Washington, D.C.: U.S. Department of Agriculture.

Zulfacar, A. (1975) 'Vegetation and the urban environment'. *Proceedings of the American Society of Civil Engineers* **101**(UP1), 21-33.

Zulfacar, A. and Clark, C.S. (1974) 'Highway noise and acoustical buffer zones'. *Proceedings of the American Society of Civil Engineers* **100** (TE2), 389-401.

6 Perspective: Environment and its control

Concern for man himself and his fate must always form the chief interest of all technical endeavour. Never forget this in the midst of your diagrams and equations [Albert Einstein]

6.1 Introduction

Previous chapters have identified the role of vegetation in a variety of situations and have endeavoured to demonstrate how its properties can be appraised and incorporated into design. Generally, a fairly restricted point of view has been pursued — both in terms of the problems examined, and in terms of the vegetation itself. Vegetation is a vital component of the biosphere and landscape; it should always be seen in this setting and as part of the overall environment. Many professionals involved in re-shaping the landscape may claim that they always maintain this broad overview, and that it represents part of their normal judgement when considering land development projects. What is 'seen', however, depends on the knowledge of the practitioner, his perception and, particularly, his viewpoint. For example, if an ecologist, a landscape architect and a hydrologist were separately to contemplate the consequences of a proposed deafforestation programme, it is highly probable that they would arrive at differing conclusions. What is important, is that the practitioner should be conscious of the limits of his perception and, additionally, should focus on the broad implications of his actions. The opening quotation by Albert Einstein epitomises this theme and reflects the thrust of this closing chapter; this aims to provide greater insight to engineers and others, into the potential impact of proposed developments and operations on the landscape and environment.

6.2 The engineer's social responsibility

An immediate question is, 'Why should the engineer concern himself with such matters?' This poses moral issues and questions professional ethics. These dictate that the engineer must always endeavour to attain and maintain the highest professional standards in serving his client and the community at large. Environmental matters are of serious concern to the public, as well as to various pressure groups, and not least to engineers themselves through the development and maintenance of pollution standards. It is not feasible for the engineer to divorce himself from this reality without compromising the ethical standards of his profession.

Engineering creations directly affect people. In all projects, engineers should learn what people need, prefer, and will tolerate; this awareness should be reflected in their solutions since the obligation to serve (well) those directly affected by their creations is an important professional and social responsibility. Engineers in general, and civil engineers in particular, like to believe that they have had a favourable impact on the environment over the years. But from the public's point of view, engineers are often the designers of systems which spoil the environment. Such a view is often misdirected, since it is public officials and business executives who make many of the decisions for which 'the engineers' are blamed. Nevertheless, it is in the self-interest of engineers to dispel these public doubts by whole-hearted acceptance of the responsibility for maintaining the quality of the environment (Stanley, 1972).

While one may pursue the case for social and environmental responsibility as a theoretical philosophical argument, it should be stressed that these principles are often embodied in legislation – as well as within the policy guidelines of the engineering profession. For example, in 1972 the American Society of Civil Engineers (A.S.C.E.) adopted an environmental policy statement expressing professional awareness and regard for the ecological balance. Its preamble states:

Environmental goals must emphasize the need for assuring a desirable quality of life in the context of expanding technology to sustain and improve human life. Implicit in these goals is the need to develop resources and facilities to improve the environment of man and the need to abate deleterious effects of technology on the environment.

The statement goes on to recommend that the civil engineer recognises the effects of his efforts upon the environment by:

(a) sharpening his awareness and increasing his knowledge and competence regarding ecological considerations;

(b) informing his client of the environmental consequences of services requested or designs selected and recommending only responsible courses of action;

(c) ceasing his services in the event that the client insists on a course of action which can be demonstrated to have undesirable consequences to the environment that outweigh benefits;

(d) weighing social and national considerations and alternatives, when appropriate, in addition to the economic and technical aspects of the project;

(e) utilising fully the mechanisms within A.S.C.E. to support his individual efforts to implement environmental decisions;

(f) recognising the urgent need for legislation and enforcement to protect the environment;

(g) providing leadership in the modification or support of governmental programmes to ensure adequate environmental protection.

The matter of a civil engineer's social responsibility has also been thoroughly debated by the Council of the Institute of Civil Engineers in the United Kingdom and culminated in a decision to remind members of their obligations in a pronouncement which was published in October 1973. The thrust of this statement is similar to the policy adopted by A.S.C.E.

6.3 Landscape and land use planning

Historically the environment has been abused; this has arisen partly because economic goals and the exploitation of technological advances have been considered more desirable than the ecological disadvantages; and partly as the result of ignorance. This attitude and its consequences, together with the effects of over-population, has resulted in the despoliation of much valuable land. If further environmental destruction is to be avoided, the exploitation of natural resources must be controlled. Planning — particularly for land use and landscape planning — is vital for human survival. Environmental damage knows no natural or territorial boundaries and therefore requires cooperation at all levels if its degradation is to be averted. Nature and wildlife can tolerate minor environmental damage, but cannot contend with the major damage arising from the pressures of population together with economic and technical development.

Lovejoy (1979) describes land as being the most vulnerable of natural resources. It is easily damaged and difficult to regenerate. Overgrazing and indiscriminate tree felling have produced serious erosion problems in many parts of the world; desertification is rampant, arising largely from man's destruction of the vegetative cover; salination ruins extensive areas of irrigated soils. These problems, and others, are examples of misuse of land involving plants. They represent processes which are difficult to remedy but which, with sound education and planning, are more easily prevented.

Land conservation should be a priority in land use policies, and in order to conserve land it is essential to secure its maximum potential use. This can be achieved by aiming at multiple uses whereby several compatible activities, e.g. recreation and grazing can be carried out on the same tract of land. Such an approach requires careful planning and implies a holistic approach, wherein no single component should be viewed in isolation.

6.3.1 Conservation

The term conservation is still popularly misidentified with naturalists protecting butterflies etc., whereas it is more appropriate to consider conservation as representing a form of resource management. It is a philosophy which is directed at the manner and timing of resource use and is now directed at the protection of environmental quality and the dignity of life. Conservation requires a fusion of approaches – political, economic, technological, ethical, aesthetic, ecological, social, and so on.

Nature conservation should be an integral part of a land use strategy for every part of the environment. Conservationists have been seeking during recent years to create amongst those responsible for the management of land and water an awareness of the practical contributions they can make to nature conservation. For example, in the United Kingdom, the publication *Conservation and Land Drainage Guidelines* (Water Space Amenity Commission, 1980) attempts to alert engineers and others to the wider issues of land drainage and represents an innovative approach to planning. Water authorities in the United Kingdom are obliged by section 22 of the 1973 Water Act to take into account the effect of their works on wildlife and conservation – as are all statutory public bodies. River management offers some of the best opportunities for retaining natural features and wildlife habitats, as illustrated in figs. 6.1 and 6.2. Figure 6.1 shows

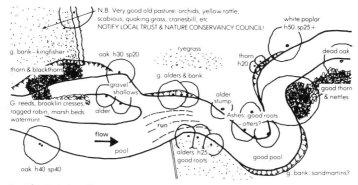

1 The Existing River

In an important spot such as this a detailed survey is needed and a survey drawing at 1:500. The Nature Conservancy Council and the local Conservation Trust will help – they should be asked early on.

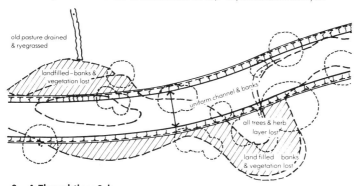

2 A Thoughtless Scheme

No effort has been made to keep existing habitats or create new ones. Canalisation produces a river with neither visual nor natural value.

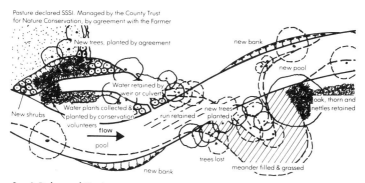

3 A Balanced Design

A reasonable compromise has been reached between agriculture and conservation. Some existing habitats are kept and new ones created.

Fig. 6.1 CONSERVATION DESIGN IN PRACTICE
Subtle alterations can benefit conservation without undue cost; new habitats can be made on unproductive land; thorough survey and early consultation are vital

that, although the straightening of a river may be the ideal for improving the ability of a channel to maintain an increased flow, it removes meanders which provide diversity in the landscape, not only for man but also for wildlife. Trees and shrubs are often removed to permit access of excavating machinery. However, the retention of trees on at least one side greatly benefits nature conservation.

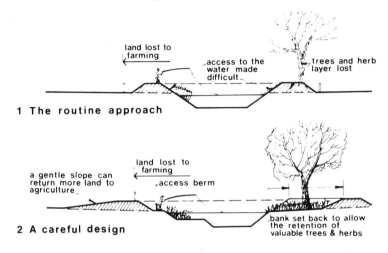

Fig. 6.2 THE DESIGN OF FLOOD BANKS
Subtle alterations in design can save or create places for wildlife and improve access to the water; such refinements take little extra land and are a necessary part of schemes; careful survey is needed to include such features at the design and budgeting stages

6.3.2 *Identifying impacts*
In many instances major projects supported by public funds will have passed through formal planning procedures in which the environmental aspects of proposed actions will have been explored. For example, in 1969 the American Congress passed the National Environmental Policy Act (N.E.P.A.) which made mandatory the preparation of an Environmental Impact Statement (E.I.S.) for any proposed federal development likely to have a significant effect on the human environment. This form of analysis makes it necessary to define environmental objectives and to develop quantitative and qualitative criteria which the designer must take into account if the project is to have a reasonable hope of approval. Although N.E.P.A. applies to projects supported by federal agencies, the majority of states within the U.S. have produced their own E.I.S. procedures. Many states require environmental impact statements for public

projects. In some states, Acts are more extensive and cover private developers, although there are considerable differences in the degree to which private developers are affected.

The E.I.S. procedure in the United States involves a fairly extensive investigation which does not necessarily lead to any formal approval of the particular project or programme. However it represents a mechanism through which public interests may be enforced and its form and content is open to challenge in the courts. In the United Kingdom, the system is somewhat different, because, in addition to statutes dealing with environmental health, there exists a statutory system of land use planning applying over the whole country and to virtually every kind of development. Within this system planning applications can be dealt with as rigorously as the planning authority thinks fit.

From this preamble it is evident that safeguards for maintaining environmental quality often rest with the legislature and are removed from the hands of the individual. For an engineer involved in such activities, it is likely that he will not be greatly involved in assessing the potential impact of his actions on the environment because this aspect will be dealt with by others. The engineer, however, can find himself in situations in which there is not this basic safeguard. For instance, in England and Wales the water authorities may proceed with river works without approaching local government for planning permission, and thus escape the usual checks applied by the planning authorities. The water authority or the individual engineer must then shoulder the responsibility of satisfying conservation interests and maintaining environmental quality. One of the main problems is knowing what to look for. Sometimes the factors will be easily identifiable because of the potential conflicts which are aroused. Examples of landscape conflicts in the United Kingdom have been tabulated by the Countryside Review Committee (1976) (see table 6.1) and underline the tension between development projects and conservation interests. While table 6.1 casts the potential impact in general terms, there is a need for a more specific and systematic check of the likely impacts. In this respect the principles embodied in the E.I.S. represent a useful guide. While no method can identify and predict all the impacts of a proposed development, the procedures proposed in Department of the Environment (1976) provide a comprehensive approach of general applicability.

The assessment of specific impacts is undertaken using the 'Technical Advice Notes' and 'List of Questions' appended to the DoE Research Report. The 'Technical Advice Notes' contains details

Table 6.1   Possible conflicts which may arise when the need to promote economic development clashes with the desire to conserve and to enhance our landscape heritage

Policy under promotion	Actions which may be taken in consequence (some with grant-aid from the Exchequer)	Policies in potential conflict
Increased food production and productivity	Ploughing, liming and fertilising	Conservation of scene, wildlife or remains; access to open country; pollution control
	Drainage work	Wildlife and water conservation
	Hedgerow clearance	Conservation of scene and wildlife
	Use of fertilisers, pesticides and herbicides	Conservation of wildlife; water quality; pollution control
	Farm and storage buildings	Conservation of scene
	Woodland clearance	Conservation of scene and wildlife
	Fencing	Conservation of scene; access to open country
	Bracken removal	Conservation of scene
	Straw-burning	Conservation of scene and wildlife; pollution control
	Free-running bulls	Access (along footpath)
Increased timber production and productivity	Afforestation and fencing	Conservation of scene; access
	Block felling	Conservation of scene
Increased piped water supply	Creation of reservoirs	Conservation of scene and productive land
Improved water quality	Groundwater schemes and estuarial storage bunds	Conservation of scene and wildlife
	Purification works	Conservation of scene
Prevention of damage by water	River training and flood protection	Conservation of scene; access
Increased mineral production	Excavation and spoil tips	Conservation of productive land; conservation of scene and wildlife; water quality; pollution control
	Transportation of materials	Conservation of amenity
National defence	Maintenance of training grounds	Food, mineral and timber production; conservation of scene; access
	Testing equipment and training	Conservation of amenity

Table 6.1   Cont'd.

Policy under promotion	Actions which may be taken in consequence (some with grant-aid from the Exchequer)	Policies in potential conflict
Promotion of tourism	Erection of catering, accommodation and service facilities	Conservation of scene, amenity and productive land
	Holiday traffic (vehicles and people)	Conservation of amenity and wildlife
Improved accessibility	New roads; building; widening and straightening of roads	Conservation of scene, wildlife and amenity; production of food
Improved housing	Take-up of countryside for new development	Conservation of productive land; conservation of scene, wildlife and amenity
Promotion of industrial development	Take-up of countryside for new factories	Conservation of productive land; conservation of scene; pollution control
Promotion of the recreational use of the countryside	Improving access to open country, conserving and protecting the footpaths system	Production of food and timber; conservation of scene and wildlife
	Providing of more facilities (country parks, picnic sites, etc.)	Production of food and timber; conservation of scene and wildlife
Clearing derelict land	Earth-moving and creative landscaping	Conservation of wildlife
Conserving the amenity of the countryside	Designation of special areas, control of development and noise, dust and other nuisance	Improvement of accessibility; housing and industrial development; tourism, recreation; extraction of minerals
Conserving the scene	Designation of special areas, control of development and noise, dust and other nuisance	Hedgerow and woodland clearance; afforestation; development generally
Conserving wildlife	Designation of special areas, control of development and noise, dust and other nuisance	Hedgerow and woodland clearance; afforestation; development generally
Improved telecommunications	Erection of masts and aerials	Conservation of scene
Increasing energy supply	Extraction of oil, gas, coal	Conservation of productive land, scene, wildlife and amenity; pollution control
	Generation and transmission of electricity	Conservation of scene
Reducing rural depopulation	Promotion of industrial development	Conservation of scene and amenity

Table 6.2   Questions to aid the assessment of: (a) impacts on land use and landscape; (b) ecological impacts

(a) Land use and landscape	(b) Ecological characteristics of the site and its surroundings
(i) Is the proposed development compatible with surrounding land uses, such as agriculture, forestry, recreation?	(i) Are the development and the existing habitats compatible?
(ii) Will the proposed development substantially alter the landscape quality of the area?	(ii) If 'yes' what conservation methods will be necessary to protect the habitats?
(iii) Will the proposed development have a substantial zone of visual influence?	(iii) 1. If the developer described conservation methods that will be used to protect sensitive habitats, are these likely to be successful? 2. Are the claims of the developer with respect to these conservation methods realistic?
(iv) How far are existing land uses within the zone of visual influence compatible with the character of the proposed development?	
(v) Is the scale of the proposed development compatible with that of the local landscape?	(iv) If the developer and habitats are not compatible, what communities will be at risk from: physical destruction; changes in groundwater level; changes in quality of standing or flowing water, oxygen content, salinity, turbidity, flow rate and temperature; chemical pollution, eutrophication and specific toxins; changes in silting pattern; air pollution of both water bodies and terrestrial habitats; dust pollution; changes in nutrient status of terrestrial habitats; opening up of other areas to increased recreation pressure by the construction of access routes, roads and pathways?
(vi) Are there any trees or buildings on the site worthy of preservation?	
(vii) Are the materials to be used in the permanent structure and buildings of the development in character with those of the local area?	
(viii) Are the landscaping proposals submitted by the applicant satisfactory?	
(ix) Has consideration been given to a satisfactory scheme for site restoration should the proposed development cease operation? Has an appropriate means of financing the implementation of the restoration scheme been agreed should the company cease to be a viable concern?	
	(v) In each of the above cases what is the local, regional and national status of any habitats at risk?
	(vi) What is the quality of the habitats regardless of status?
	(vii) What dependent habitats or communities will also be at risk including non-residents and migrants? What is their status?
	(viii) Can any of these habitats be recreated within a short period (5-10 years)?

of techniques for assessing impacts, e.g. noise, the zone of visual influence, etc. The 'List of Questions' highlights key aspects of various impacts which might be associated with a development; the questions are not exhaustive and may require modifications for specific objectives. Questions to aid assessment of ecological impacts and impacts of landscape character and land use are listed in table 6.2. The DoE report goes on to show how a formal impact analysis can be produced. The value of approach lies not so much in the formal framework which is developed, but in the questions themselves. These can alert the engineer to contentious aspects of his work; also, they may suggest ways in which a design can be modified to reduce associated conflict, and to enhance its harmony with the natural environment (see figs. 6.1 and 6.2).

There are, of course, many recipes for producing Environmental Impact Statements — a useful review being given in Clark *et al.* (1979). Cheremisinoff and Morresi (1977) give a valuable introduction to the preparation of an impact statement and analyse various developments likely to be encountered by engineers; this approach, together with texts such as Lovejoy (1979) and Beatty *et al.* (1979), provides a balanced view of land use and landscape planning.

6.4 References

Beatty, M.T., Peterson, G.W., and Swindale, L.D. *eds.* (1979) *Planning the Uses and Management of Land*. Agronomy No. 21. Madison: American Society of Agronomy and Crop Science Society of America and Soil Science Society of America Inc.

Cheremisinoff, P.N. and Morresi, A.C. (1977) *Environmental Assessment and Impact Statement Handbook*. Michigan, U.S.: Ann Arbor Science.

Clark, B.D., Chapman, K., Bisset, R. and Wathern, P. (1979) 'Environmental impact analysis'. In Lovejoy, D. *ed. Land Use and Landscape Planning*. (2nd edn.) Glasgow: Leonard Hill, 51-87.

Countryside Review Committee (1976) *The Countryside — Problems and Policies*. London: Her Majesty's Stationery Office.

Department of the Environment (1976) *Assessment of Major Industrial Applications: A Manual*. D.o.E. Research Report. No. 13. London.

Lovejoy, D. (ed.) (1979) *Land Use and Landscape Planning*. 2nd edn. Glasgow: Leonard Hill.

Stanley, C.M. (1972) 'The engineer and the environment'. *Civil Engineering-ASCE* **42** (July), 78-79.

Water Space Amenity Commission (1980) *Conservation and Land Drainage Guidelines*. London: Water Space Amenity Commission.

Appendix 1
Water Potential

Following the introduction given in section 1.3, information is presented to clarify the definition of water potential in plants and soils; its units; and its value in discerning the status and movement of water in the soil-plant system.

Soil water potential

At equilibrium, the constituent water of the liquid phase is subject to the action of the gravitation field, the influence of dissolved salts and of the solid phase (including adsorbed ions) in its given geometry of packing, and to the action of the local pressure in the soil gas phase. Together these factors determine the value of the total potential ψ_t of the constituent water relative to a chosen standard state. The standard state is generally considered to be a pool of pure water (i.e. no dissolved salts), free water (i.e. water not influenced by the solid phase) at a chosen temperature, height and atmospheric pressure.

With this concept, the total water potential refers to the useful work per unit mass of pure water that must be done to transfer a small unit of water from the standard state to the soil-liquid phase at the point under consideration.

By dividing this transfer into a sequence of carefully defined steps, each reflecting a differing force field, the total potential may be found from the relation:

$$\psi_{\text{soil}} = \psi_g + \psi_o + \psi_p \qquad\qquad [A1.1]$$

where ψ_g is the gravitational potential, ψ_o the osmotic potential and ψ_p the pressure potential.

Gravitational potential

The gravitational potential is the work done against the gravitational field in displacing water from the standard state to a position at the same height by $\psi_g = \rho g \Delta h$ in which g is the gravitation acceleration; ρ, the density of water; and Δh, the difference in height between a standard reference plane and the position in the soil system.

Osmotic potential

The osmotic potential, which refers to the effect of solutes, may be defined as having zero potential under standard conditions and is otherwise negative; this arises because the addition of a solute reduces the potential. The negative sign may also be interpreted as signifying 'solute suction', because osmotic pressure causes a net displacement of water molecules from the standard pool towards the point under consideration in the soil solution.

Pressure potential

Whereas the components ψ_g and ψ_o are determined by height and osmotic pressure in the soil solution, the pressure potential is determined by the water content of the soil.

In saturated conditions the water in the soil will be at a pressure exceeding atmospheric pressure due to the static head. Since the reference pressure in the standard pool is taken as atmospheric pressure, it implies that work must be carried out against the pressure gradient during the transfer from the standard state to the position under consideration. If atmospheric pressure is at its standard value, ψ_p is positive.

In unsaturated conditions water is subject to capillary forces and surface adsorption. Because these forces attract water molecules towards the solid phase, the resultant potential is negative; this is referred to as the matric potential ψ_m, provided the atmospheric pressure is at its standard value.

In dry soils, adsorption largely determines the matric potential whereas in moist soils, water retention is primarily due to the surface tension forces which develop in liquid/air interfaces. As pore sizes become smaller, the surface tension forces become more significant and the matric potential becomes increasingly negative. This implies for a given water content, that it is more difficult to extract water held under tension from a fine textured soil than from one with a coarser texture.

Suction

For convenience of dropping the negative sign, the matric potential is regarded as being equivalent to a matric suction of magnitude $|\psi_m|$ (>0). Similarly, osmotic pressure can be regarded as a solute suction which when added to the matric suction yields the total suction.

Units

The unit quantity of water on which potential is based can be a volume, mass or weight. Potential per unit volume ($J\ m^{-3}$) has units equivalent to pressure. Potential per unit mass of water is expressed as ($J\ kg^{-1}$) and per unit weight is a length (head) in metres. A potential per unit weight of h(m) is equivalent to gh ($J\ kg^{-1}$) on a mass basis or ρgh ($N\ m^{-2}$) on a volume basis. In these conversions, g can be taken as $9.81\ m\ s^{-2}$ and the density of water ρ as $1000\ kg\ m^{-3}$ with sufficient accuracy for most purposes.

pF

Because the soil moisture tension can vary over several orders of magnitude between wet and dry state, a convenient unit is pF = $\log_{10} 100\ |h|$ where $|h|$ is tension expressed in metres of head equivalent.

Potential and plant cells

In plant cells the water potential ψ_{cell} may be expressed by the approximate equilibrium relationship:

$$\psi_{cell} = \psi_o + \psi_p \qquad\qquad [A1.2]$$

in which ψ_o expresses the effect of solutes in the cell solution (osmotic potential) and ψ_p the effect of turgor pressure (pressure potential). Turgor pressure is almost always positive whereas ψ_s is negative. The sum of the terms is a negative number (i.e. held under tension) except in fully turgid cells when it becomes zero. For example, if $\psi_o = -20$ bar the relationship for cells can be depicted as follows using equation [A1.2]

	ψ_{cell}	=	ψ_o	+	ψ_p	
Turgid	0	=	−20	+	(+20)	bar
Partly turgid	−10	=	−20	+	(+10)	bar
Flaccid	−20	=	−20	+	(0)	bar

From this example it can be seen that when the turgor pressure is low (as in a situation of water stress) the cells become flaccid and the plant may wilt.

Water movement

One of the main uses of water potential is to ascertain the direction and volume of water flow between any two points. For plants in contact with soil moisture, only the difference between ψ_{soil} and ψ_{plant} is vital. As long as $\psi_{soil} > \psi_{plant}$ water will continue to flow into the roots and will continue until the pressures become equal. This will come about by gradual changes of pressure potential when transpirational vapour loss slows down. The following numerical example from Meidner and Sheriff (1976) illustrates this behaviour:

Transpiring plant	Soil	Non-transpiring plant
$\psi_o = -3$ bar $\psi_p = -7$ bar		$\psi_o = -3$ bar $\psi_p = +1$ bar equal to the difference between ψ_{soil}
	$\psi_{soil} = -2$ bar	and ψ_o
$\psi_{plant} = -10$ bar		$\psi_{plant} = -2$ bar
$\psi_{plant} - \psi_{soil} = -8$ bar:		$\psi_{plant} - \psi_{soil} = 0$:
therefore net inflow of water into plant		therefore no net inflow of water into plant

Gradients in water potential not only determine the direction of flow, but are instrumental in determining its magnitude. In many respects the process of water movement in the soil-plant-air system may be considered as analogous to the flow of an electric current. This assumption implies that the equivalent of Ohm's law can be applied to determine the volume of water moving between two points i.e.,

$$Q = \frac{\Delta\psi}{R} \qquad\qquad [A1.3]$$

where Q is the volume of water flowing per unit time, $\Delta\psi$ the potential difference between the two points and R the resistance to flow between the points. For example, Rutter (1975) reported that the rate of uptake per unit area by plant roots can be determined by:

$$Q = \frac{\psi_{soil} - \psi_{leaf}}{R_{soil} + R_{plant}}$$
[A1.4]

in which R_{soil} signifies the overall hydraulic resistance to movement of liquid from the bulk soil to the root system and R_{plant}, the resistance from the root surfaces to the cells of the leaf. This is a simple representation of a situation which is very complex, but nevertheless illustrates an important analytical technique for discerning water movement in the soil-plant-air system. The success of the approach depends on the evaluation of the determining parameters.

References

Meidner, H. and Sheriff, H.W. (1976) *Water and Plants*. Glasgow: Blackie.
Rutter, A.J. (1975) 'The hydrological cycle in vegetation'. In: *Vegetation and the Atmosphere*. (Monteith, J.L., ed.). London: Academic Press, 111-154.

Appendix 2
Species Index

Common name	Botanical name
Alfalfa	*Medicago sativa*
Alder	*Alnus* spp.
American beach grass	*Ammophila breviligulata*
American elm	*Ulmus americana*
Ash	*Fraxinus excelsior*
Aspen	*Populus tremula*
Barley	*Hordeum vulgare*
Beans (field)	*Phaseolus vulgaris*
Beet	*Beta vulgaris*
Bermuda grass	*Cynodon dactylon*
Birch	*Betula* spp.
Black locust	*Robinia pseudoacacia*
Black poplar	*Populus nigra*
Blackthorn	*Prunus spinosa*
Blue lupin	*Lupinus angustifolius*
Brown top	*Agrostis tenuis*
Cabbage	*Brassica oleracea capitata*
Carrot	*Daucus carota*
Citrus	*Citrus* spp.
Clover	*Trifolium* spp.
Coast Douglas fir	*Pseudotsuga menziesii*
Collards	*Brassica oleracea* var. *acephala*
Common sallow	*Salix cinerea*
Corn	*Zea mays*
Cottonwood	*Populus* spp.
Cranesbill	*Geranium* spp.
Diamond willow	*Salix* sp.
Douglas fir	*Pseudotsuga* spp.
Elm	*Ulmus* spp.

Goat willow	*Salix caprea*
Gorse	*Ulex* spp.
Hard fescue	*Festuca longifolia*
Hickory	*Carya* spp.
Italian rye-grass	*Lolium multiflorum*
Kale	*Brassica oleracea* var. *acephala*
Kudzu vine	*Pueraria thunbergiana*
Lettuce	*Lactuca sativa*
Love grass	*Eragrostis minor*
Lupin	*Lupinus* spp.
Lyme grass	*Elymus arenarius*
Marram grass	*Ammophila arenaria*
Millipede	*Diplopoda*
Mustards	*Brassica* spp.
Nettle	*Urtica* spp.
Oak	*Quercus* spp.
Oats	*Avena sativa*
Orchids	*Orchidaceae*
Otter	*Lutra*
Perennial rye-grass	*Lolium perenne*
Poplar	*Populus* spp.
Quaking grass	*Briza* spp.
Red fescue	*Festuca rubra*
Reed canary-grass	*Phalaris arundinacea*
Rocky Mountain Douglas fir	*Pseudotsuga menziesii* var. *glauca*
Sallows	*Salix* spp.
Salt grass	*Distichlis stricta*
Sand couch	*Agropyron junceiforme*
Scabious	*Scabiosa* spp.; *knautia* spp.; *Succisa* spp.
Sedges	*Cyperaceae*
Siberian elm	*Ulmus pumila*
Soybeans	*Glycine max*
Sycamore	*Acer pseudoplatanus*
Tall fescue	*Festuca arundinacea*
Tall wheatgrass	*Agropyron elongatum*

Thorn	*Crataegus* spp.
Timothy-grass	*Phleum pratense*
Tomato	*Lycopersicon esculentum*
Turnip	*Brassica rapa*
Watermint	*Mentha aquita*
Wheat	*Triticum aestivum*
White clover	*Trifolium repens*
White poplar	*Populus alba*
Willow	*Salix* spp.
Yellow rattle	*Rhinanthus minor*
Yellow willow	*Salix lutea*

Appendix 3
Conversion of Weights and Measures

	Unit	Equivalent unit
Length	1 m	39.37 in
		3.281 ft
	1 km	0.6214 mile
Area	1 m^2	1550 sq. inch
		10.76 ft^2
		1.196 sq. yard
	1 ha (10^4 m^2)	2.471 acre
	1 km^2	0.3861 sq. mile
Volume	1 m^3	35.31 ft^3
Mass	1 kg	35.27 ounces
		2.204 lb
	1000 kg	1.102 ton (2000 lb)
		0.9842 ton (2240 lb)
Capacity	1 l	0.2642 gallons (US)
		0.2200 gallons (Imperial)
	1 m^3	28.38 bushels (US)
		27.50 bushels (Imperial)
Pressure	1 N m^{-2}	0.01 mbar
		0.145 $\times$ 10^{-3} p.s.i.
Heat energy	1 J	0.2388 cal
Heat or radiation flux	1 W (= 1 J s^{-1})	0.2388 cal s^{-1}
Radiation flux density	1 W m^{-2}	2.388 $\times$ 10^{-5} cal cm^{-2} s^{-1}
Latent heat	1 J kg^{-1}	2.388 $\times$ 10^{-4} cal g^{-1}
Specific heat	1 J kg^{-1} °C^{-1}	2.388 $\times$ 10^{-4} cal g^{-1} °C^{-1}

Temperature	$1°C$	$1.8°F$
Application rate	$1\ l\ ha^{-1}$	0.1069 gallons $acre^{-1}$ (US)
	$1\ kg\ m^{-2}$	29.49 oz $yard^{-2}$
	$1\ kg\ ha^{-1}$	0.8922 lb $acre^{-1}$
	$1\ t\ ha^{-1}$	0.4047 t $acre^{-1}$

Rainfall kinetic energy $1\ m\ t\ ha^{-1}\ cm^{-1}$ 0.269 ft t $ha^{-1}\ in^{-1}$

Concentration

$$1\ meq\ Na^+\ l^{-1} = \frac{1}{22.99}\ mg\ Na^+\ l^{-1}$$

$$1\ meq\ Ca^{2+}\ l^{-1} = \frac{1}{20.04}\ mg\ Ca^{2+}\ l^{-1}$$

$$1\ meq\ Mg^{2+}\ l^{-1} = \frac{1}{12.15}\ mg\ Mg^{2+}\ l^{-1}$$

$$1\ meq\ Zn^{2+}\ l^{-1} = \frac{1}{32.69}\ mg\ Zn^{2+}\ l^{-1}$$

Acknowledgements and Sources

We acknowledge the following sources and thank the quoted publishers and individuals for granting permission to reproduce material; full references are listed at the end of the corresponding chapters.

Chapter 1

Fig. 1.1 from Winter (1974) — Macmillan, London and Basingstoke; fig. 1.2 from Etherington (1975) — John Wiley and Sons, New York; fig. 1.3 from Buckman and Brady (1969) — Macmillan Publishing Co. Inc., New York; fig. 1.4 adapted from Mehlich (1941) — Soil Science Society of America, Madison; fig. 1.5, table 1.2 and example on page 12 from Thompson and Troeh (1973) — McGraw Hill Book Company, New York; fig. 1.7 adapted from Truog (1946) — Soil Science Society of America, Madison; fig. 1.8 from Young (1968) — Soil Conservation Society of America, Ankeny; table 1.3 from Sopher and Baird (1978) — Reston Publishing Company, a Prentice-Hall Co., Reston VA: table 1.4 from Bradshaw and Chadwick (1980) — Blackwell Scientific Publications, Oxford and Professor A D Bradshaw.

Chapter 2

Fig. 2.4 from Calder (1979) — Fuel and Metallurgical Journals Ltd., Redhill, U.K.; fig. 2.5 from Hillel (1971) — Academic Press, New York; fig. 2.7 adapted from Strahler (1975); fig. 2.9 from Rantz (1971); fig. 2.10 adapted from Ogrosky and Mockus (1964) — McGraw-Hill Book Company, New York; fig. 2.11 adapted from Dunne and Leopold (1978) — copyright © (1978) by W.H. Freeman and Company, New York; fig. 2.13 adapted from Doorenbos and Pruitt (1977) — Food and Agriculture Organisation of the United

Nations, Rome; table 2.3 adapted from Salter and Williams (1965) as in Rutter (1975); tables 2.4 and 2.5 from Grindley (1969) — International Association of Hydrological Sciences, Wallingford, U.K.; table 2.9 adapted from Holtan and Lopez (1971); table 2.10 adapted from Wiesner (1970); table 2.11 from American Society of Civil Engineers (1949) — American Society of Civil Engineers, New York; tables 2.12, 2.13, 2.15 and 2.16 from U.S. Soil Conservation Service (1972); table 2.14 based on Musgrave and Holtan (1964); table 2.17 from Chow — McGraw Hill Book Company, New York; table 2.18 from Bell and Songthara om Kar (1969) — Elsevier Scientific Publishing Company, Amsterdam; tables 2.21 and 2.22 from Chow (1959) — McGraw Hill Book Company, New York; table 2.23 from Water Space Amenity Commission (1980) — Water Space Amenity Commission, London.

Chapter 3

Fig. 3.1 redrawn from Wolman (1967) — Svenska sällskapet för antropolgi och geografi, Stockholm; fig. 3.2 from Gray (1978) — The New South Wales Institute of Technology, and the University of New South Wales; fig. 3.3 adapted from Wu et al. (1979) — National Research Council of Canada; figs. 3.4 to 3.6 from Rice (1977) — Food and Agriculture Organisation of the United Nations, Rome; fig. 3.7 from Wischmeier et al. (1976) — Soil Conservation Society of America, Ankeny; fig. 3.8 from Meyer and Kramer (1969) — American Society of Agricultural Engineers, St Joseph; fig. 3.9 adapted from Kohnke (1950) — Academic Press, New York; fig. 3.10, table 3.7 and table 3.9 from Bradshaw and Chadwick (1980) — Blackwell Scientific Publications, Oxford and Professor A D Bradshaw; figs 3.12 to 3.14 and 3.17 from Schiechtl (1980) — The University of Alberta Press, Edmonton; fig. 3.15 and major extracts in section 3.11 from Bache and MacAskill (1981) — Elsevier Scientific Publishing Co., Amsterdam; fig. 3.16 from Seibert (1968) — Council of Europe, Strasbourg, France; fig. 3.19 redrawn from U.S. Department of Agriculture (1975); fig. 3.20 redrawn from Woodhouse (1978) — U.S. Army Coastal Engineering Research Center; fig. 3.21 from Boorman (1977) — John Wiley & Sons Ltd; table 3.2 from Fleming (1969) — The Institution of Civil Engineers, London; tables 3.3 to 3.5 from U.S. Soil Conservation Service (1975); table 3.10 adapted from U.S. Soil Conservation Service (1954); equation [3.11] — presentation as in fig. 2.2 of Hudson (1971) — Batsford, London.

Chapter 4

Fig. 4.1 and 4.2 from Overcash and Pal (1979) — Butterworths, Sevenoaks, U.K.; fig. 4.3 adapted from U.S. Environmental Protection Agency (1977); figs. 4.4 and 4.5 from Loehr *et al.* (1979) — Van Nostrand Reinhold, New York; table 4.2 from Spyridakis and Welch (1976) — Butterworth, Sevenoaks, U.K.; table 4.3 adapted from Carlile and Phillips (1976); table 4.5 from Davis and Carlton-Smith (1980) — the Water Research Centre, Medmenham, U.K.; table 4.6 from Dowdy *et al,* (1976) — Soil Conservation Society of America, Ankeny.

Chapter 5

Fig. 5.1 adapted from Plate (1971) in the form shown by Businger (1975); figs 5.2 and 5.7 after Nägeli (1946); fig. 5.3 adapted from Seginer (1975) — D. Reidal Publishing Company, Dordrecht, Holland; fig. 5.4 based on Caborn (1965); fig. 5.5 after Bates (1944); fig. 5.8 from Caborn (1965) — Faber and Faber Ltd, London; fig. 5.10 from Skidmore (1965) — Soil Science Society of America, Madison, WI; figs. 5.11 and 5.12 from Skidmore and Hagen — American Society of Agricultural Engineers, St Joseph, MI; fig. 5.13 from Marshall (1967) — Commonwealth Agricultural Bureau, Slough, U.K.; figs. 5.14 and 5.15 from Frank and George (1975) — Great Plains Agricultural Council, Lincoln, U.S.A.; fig. 5.16 from Scholes and Sargent (1971) — Building Research Establishment and the Controller, Her Majesty's Stationery Office, Crown copyright; fig. 5.17 from Organisation for Economic Cooperation and Development (1971) — O.E.C.D., Paris; fig. 5.18 based on Caborn (1965); fig. 5.19 adapted from Robinette (1972); fig. 5.20 based on Caborn (1965); fig. 5.21 after Kühlewind *et al.* (1955).

Chapter 6

Figs. 6.1 and 6.2 from Water Space Amenity Commission (1980) — Water Space Amenity Commission, London; table 6.1 from Countryside Review Committee (1976) — Controller of Her Majesty's Stationery Office, Norwich; table 6.2 from Department of the Environment (1976) — Controller of Her Majesty's Stationery Office, Norwich.

Appendix 1

Numerical example on page 291 is taken from Meidner and Sheriff (1976) – Blackie and Sons Ltd., Glasgow.

Index

Abrasion (*see also* scouring), 153-155
Accretion (*see also* siltation), 152,
  155-156, 159, 174-178
Acids/acidity (*see also* alkalinity, pH)
  climate and, 14
  exchange, 14, 137
  long term, 141
  nitrification and, 201
  nutrient exchange and, 18-20, 207
  sand-dunes and, 175
  soil, 13-15, 136-137, 140, 214
  vegetation and, 14-15, 214
  waste application of, 209-210, 214
Acoustic buffer zones, 251-253
Adsorption
  bases, 13, 211
  heavy metals, 216-217
  phosphorous fixation, 192, 207
  plant nutrients, 14
  process, 12, 187, 192, 195-196
  specific, 197
  water, 289
Aeration, soils (*see* aerobic/anaerobic,
  gaseous transfer)
Aerobic/anaerobic conditions
  decomposition and, 186, 194-195,
    221-222
  effects on plant, 3, 8, 163, 222
  nitrification/dentrification,
    201-202
Aerosol drift, 262-263
Afforestation (*see also* forests), 24, 96,
  97, 99, 178, 284
After-care, 138
Aggregates, 10, 11, 52, 138, 210-211
Agricultural land
  conservation/ecological aspects,
    280-82, 284, 286
  erosion of, 120, 123-124, 128-130,
    151, 241, 244
  hydrological aspects, 23, 52-54, 57-61,
    62-63, 85-93

land use, 131, 155
  waste applications on, 190-191,
    222-224
  shelterbelts, 228-229, 264-265
  water supply/snow, 247, 250 (*see
    also* irrigation)
Air pollution, 228, 254-263
Albedo, 27-28, 31-32, 73-74, 95
Alder/alderwood, 21, 143, 160-161,
  293
Alkaline soils/alkalinity (*see also*
  acidity, pH), 14, 19, 207
Aluminium, 13-14, 16, 207
Ammonification/ammonium (*see*
  nitrogen)
Anions, 16, 188, 190, 192, 195-196,
  200, 207
Application rate
  fertilisers, 140
  irrigation water, 89-93, 197-199,
    205, 211-215
  mulches/stabilisers, 150
  waste constituents (*see* assimilative
    capacity)
Aquatic plants, 157-159
Arable land (*see also* agricultural land),
  118, 120, 123
Ash, 160, 293
Assimilation of wastes/wastewater, 12,
  185-227
Assimilative capacity
  acids, bases, salts, 209-210
  bio-organic wastes, 221
  concept, 187, 189-190
  heavy metals, 215
  nitrogen, 203-204
  phosphorus, 208-209
  suspended solids, 194
  water, 198
Atmospheric stability, 258-260
Atmospheric turbulence, 257, 259
Available energy, 27, 73

302